Molecular Biology

Molecular Biology

Edited by
Curtis Holmes

Molecular Biology
Edited by Curtis Holmes
ISBN: 978-1-63549-190-6 (Hardback)

Published by Larsen and Keller Education,
5 Penn Plaza,
19th Floor,
New York, NY 10001, USA

Cataloging-in-Publication Data

Molecular biology / edited by Curtis Holmes.
p. cm.
Includes bibliographical references and index.
ISBN 978-1-63549-190-6
1. Molecular biology. 2. Molecular microbiology. 3. Biomolecules. 4. Molecular biology--Technique.
I. Holmes, Curtis.
QH506 .M65 2017
572.8--dc23

Printed and bound in the United States of America.

For more information regarding Larsen and Keller Education and its products, please visit the publisher's website www.larsen-keller.com

Table of Contents

Preface

This book is a valuable compilation of topics, ranging from the basic to the most complex theories and principles in the field of molecular biology. It provides thorough insights about the varied sub-branches of this field. Molecular biology refers to the study of the cells and the different biological activities happening between biomolecules in different cells. Also included in this text is a detailed explanation of the various concepts and applications of molecular biology. The book is a complete source of knowledge on the present status of this important field. It will be of great help to students in the field of biochemistry, genetics and bioinformation.

Given below is the chapter wise description of the book:

Chapter 1- The branch of science that concerns biological activity at the molecular level is regarded as molecular biology. This field overlaps with biology and chemistry. This chapter on molecular biology introduces the reader to the basic concepts of the subject including molecular biology, molecular microbiology and biomolecule. The chapter provides an integrated understanding on molecular biology.

Chapter 2- The processes involved in molecular biology are DNA replication, transcription, gene expression and translation. The biological process of producing two identical replicas of DNA from one original DNA molecule is referred to as DNA replication. This chapter serves as a source to understand the major categories related to molecular biology.

Chapter 3- Molecular cloning is a method used to assemble recombinant DNA molecules and polymerase chain reaction is the copying of a piece of DNA to generate millions of copies of the same DNA. This chapter discusses the techniques of molecular biology in a critical manner providing key analysis to the subject matter.

Chapter 4- The method used to imitate the behavior of molecules is molecular modelling whereas the testing of the properties of molecules in order to assemble better systems and processes is molecular engineering. This chapter is a compilation of fields like molecular modelling, molecular engineering, molecular genetics and molecular microbiology.

Chapter 5- This chapter will provide an integrated understanding of the applications of molecular biology, with a brief explanation on molecular medicine, bioinformatics and central dogma of molecular biology. This chapter helps the reader to develop a better understanding on the applications of molecular biology.

Chapter 6- Blotting can be divided into three techniques, southern blotting, northern blotting and western blotting. The macromolecule analyzed in southern blot technique is DNA, for northern blot it is the RNA and for western it is protein. The chapter on macromolecule blotting offers an insightful focus, keeping in mind the complex subject matter.

Chapter 7- The history of molecular biology began in the 1930's with the hope of understanding life better and at a fundamental level. In modern times, molecular biology pursuits to explain life from macromolecular properties that generates life. This chapter explains to the reader the significance of the history of molecular biology and the progress it has made over a period of years.

At the end, I would like to thank all those who dedicated their time and efforts for the successful completion of this book. I also wish to convey my gratitude towards my friends and family who supported me at every step.

Editor

1

Introduction to Molecular Biology

The branch of science that concerns biological activity at the molecular level is regarded as molecular biology. This field overlaps with biology and chemistry. This chapter on molecular biology introduces the reader to the basic concepts of the subject including molecular biology, molecular microbiology and biomolecule. The chapter provides an integrated understanding on molecular biology.

Molecular Biology

Molecular biology concerns the molecular basis of biological activity between biomolecules in the various systems of a cell, including the interactions between DNA, RNA and proteins and their biosynthesis, as well as the regulation of these interactions. Writing in *Nature* in 1961, William Astbury described molecular biology as:

"*...not so much a technique as an approach, an approach from the viewpoint of the so-called basic sciences with the leading idea of searching below the large-scale manifestations of classical biology for the corresponding molecular plan. It is concerned particularly with the* forms *of biological molecules and [...] is predominantly three-dimensional and structural—which does not mean, however, that it is merely a refinement of morphology. It must at the same time inquire into genesis and function.*"

Relationship to Other Biological Sciences

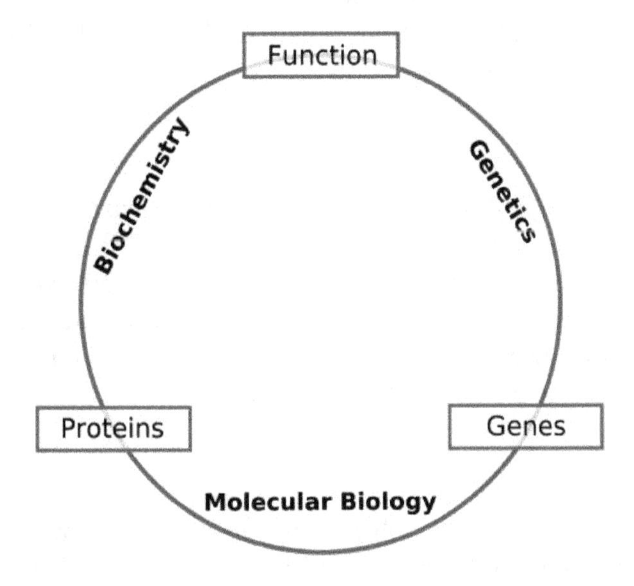

Schematic relationship between biochemistry, genetics and molecular biology

Researchers in molecular biology use specific techniques native to molecular biology but increasingly combine these with techniques and ideas from genetics and biochemistry. There is not a defined line between these disciplines. The figure to the right is a schematic that depicts one possible view of the relationship between the fields:

- *Biochemistry* is the study of the chemical substances and vital processes occurring in live organisms. Biochemists focus heavily on the role, function, and structure of biomolecules. The study of the chemistry behind biological processes and the synthesis of biologically active molecules are examples of biochemistry.
- *Genetics* is the study of the effect of genetic differences on organisms. This can often be inferred by the absence of a normal component (e.g. one gene). The study of "mutants" – organisms which lack one or more functional components with respect to the so-called "wild type" or normal phenotype. Genetic interactions (epistasis) can often confound simple interpretations of such "knockout" studies.
- *Molecular biology* is the study of molecular underpinnings of the processes of replication, transcription, translation, and cell function. The central dogma of molecular biology where genetic material is transcribed into RNA and then translated into protein, despite being an oversimplified picture of molecular biology, still provides a good starting point for understanding the field. This picture, however, is undergoing revision in light of emerging novel roles for RNA.

Much of the work in molecular biology is quantitative, and recently much work has been done at the interface of molecular biology and computer science in bioinformatics and computational biology. As of the early 2000s, the study of gene structure and function, molecular genetics, has been among the most prominent sub-field of molecular biology.Increasingly many other loops of biology focus on molecules, either directly studying their interactions in their own right such as in cell biology and developmental biology, or indirectly, where the techniques of molecular biology are used to infer historical attributes of populations or species, as in fields in evolutionary biology such as population genetics and phylogenetics. There is also a long tradition of studying biomolecules "from the ground up" in biophysics.

Techniques of Molecular Biology

Since the late 1950s and early 1960s, molecular biologists have learned to characterize, isolate, and manipulate the molecular components of cells and organisms. These components include DNA, the repository of genetic information; RNA, a close relative of DNA whose functions range from serving as a temporary working copy of DNA to actual structural and enzymatic functions as well as a functional and structural part of the translational apparatus, the ribosome; and proteins, the major structural and enzymatic type of molecule in cells.

Molecular Cloning

One of the most basic techniques of molecular biology to study protein function is molecular cloning. In this technique, DNA coding for a protein of interest is cloned (using PCR and/or restriction enzymes) into a plasmid (known as an expression vector). A vector has 3 distinctive features:

an origin of replication, a multiple cloning site (MCS), and a selective marker (usually antibiotic resistance). The origin of replication will have promoter regions upstream from the replication/ transcription start site.

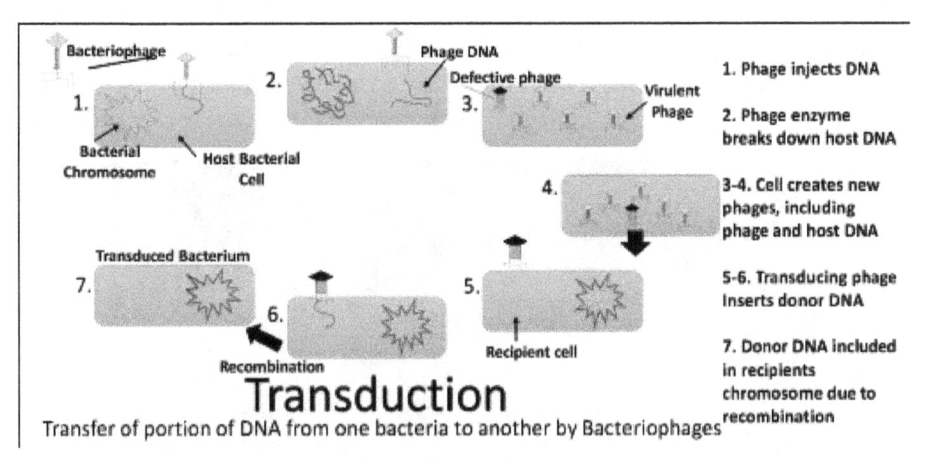

Transduction image

This plasmid can be inserted into either bacterial or animal cells. Introducing DNA into bacterial cells can be done by transformation (via uptake of naked DNA), conjugation (via cell-cell contact) or by transduction (via viral vector). Introducing DNA into eukaryotic cells, such as animal cells, by physical or chemical means is called transfection. Several different transfection techniques are available, such as calcium phosphate transfection, electroporation, microinjection and liposome transfection. DNA can also be introduced into eukaryotic cells using viruses or bacteria as carriers, the latter is sometimes called bactofection and in particular uses *Agrobacterium tumefaciens*. The plasmid may be integrated into the genome, resulting in a stable transfection, or may remain independent of the genome, called transient transfection.

In either case, DNA coding for a protein of interest is now inside a cell, and the protein can now be expressed. A variety of systems, such as inducible promoters and specific cell-signaling factors, are available to help express the protein of interest at high levels. Large quantities of a protein can then be extracted from the bacterial or eukaryotic cell. The protein can be tested for enzymatic activity under a variety of situations, the protein may be crystallized so its tertiary structure can be studied, or, in the pharmaceutical industry, the activity of new drugs against the protein can be studied.

Polymerase Chain Reaction (PCR)

Polymerase chain reaction is an extremely versatile technique for copying DNA. In brief, PCR allows a specific DNA sequence to be copied or modified in predetermined ways. The reaction is extremely powerful and under perfect conditions could amplify 1 DNA molecule to become 1.07 Billion molecules in less than 2 hours. The PCR technique can be used to introduce restriction enzyme sites to ends of DNA molecules, or to mutate (change) particular bases of DNA, the latter is a method referred to as site-directed mutagenesis. PCR can also be used to determine whether a particular DNA fragment is found in a cDNA library. PCR has many variations, like reverse transcription PCR (RT-PCR) for amplification of RNA, and, more recently, quantitative PCR which allow for quantitative measurement of DNA or RNA molecules.

Gel Electrophoresis

Gel electrophoresis is one of the principal tools of molecular biology. The basic principle is that DNA, RNA, and proteins can all be separated by means of an electric field and size. In agarose gel electrophoresis, DNA and RNA can be separated on the basis of size by running the DNA through an electrically charged agarose gel. Proteins can be separated on the basis of size by using an SDS-PAGE gel, or on the basis of size and their electric charge by using what is known as a 2D gel electrophoresis.

Two percent Agarose Gel in Borate Buffer cast in a Gel Tray (Front, angled)

Macromolecule Blotting and Probing

The terms *northern*, *western* and *eastern* blotting are derived from what initially was a molecular biology joke that played on the term *Southern blotting*, after the technique described by Edwin Southern for the hybridisation of blotted DNA. Patricia Thomas, developer of the RNA blot which then became known as the *northern blot*, actually didn't use the term. Further combinations of these techniques produced such terms as *southwesterns* (protein-DNA hybridizations), *northwesterns* (to detect protein-RNA interactions) and *farwesterns* (protein-protein interactions), all of which are presently found in the literature.

Southern Blotting

Named after its inventor, biologist Edwin Southern, the Southern blot is a method for probing for the presence of a specific DNA sequence within a DNA sample. DNA samples before or after restriction enzyme (restriction endonuclease) digestion are separated by gel electrophoresis and then transferred to a membrane by blotting via capillary action. The membrane is then exposed to a labeled DNA probe that has a complement base sequence to the sequence on the DNA of interest. Most original protocols used radioactive labels, however non-radioactive alternatives are now available. Southern blotting is less commonly used in laboratory science due to the capacity of other techniques, such as PCR, to detect specific DNA sequences from DNA samples. These blots are still used for some applications, however, such as measuring transgene copy number in transgenic mice, or in the engineering of gene knockout embryonic stem cell lines.

Northern Blotting

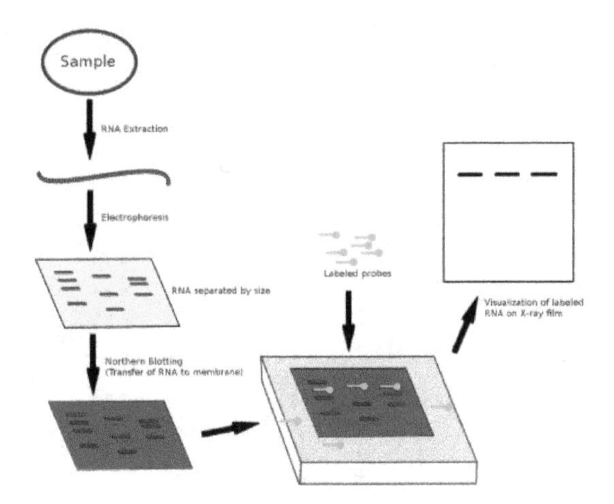

Northern blot diagram

The northern blot is used to study the expression patterns of a specific type of RNA molecule as relative comparison among a set of different samples of RNA. It is essentially a combination of denaturing RNA gel electrophoresis, and a blot. In this process RNA is separated based on size and is then transferred to a membrane that is then probed with a labeled complement of a sequence of interest. The results may be visualized through a variety of ways depending on the label used; however, most result in the revelation of bands representing the sizes of the RNA detected in sample. The intensity of these bands is related to the amount of the target RNA in the samples analyzed. The procedure is commonly used to study when and how much gene expression is occurring by measuring how much of that RNA is present in different samples. It is one of the most basic tools for determining at what time, and under what conditions, certain genes are expressed in living tissues.

Western Blotting

Antibodies to most proteins can be created by injecting small amounts of the protein into an animal such as a mouse, rabbit, sheep, or donkey (polyclonal antibodies) or produced in cell culture (monoclonal antibodies). These antibodies can be used for a variety of analytical and preparative techniques.

In western blotting, proteins are first separated by size, in a thin gel sandwiched between two glass plates in a technique known as SDS-PAGE (sodium dodecyl sulfate polyacrylamide gel electrophoresis). The proteins in the gel are then transferred to a polyvinylidene fluoride (PVDF), nitrocellulose, nylon, or other support membrane. This membrane can then be probed with solutions of antibodies. Antibodies that specifically bind to the protein of interest can then be visualized by a variety of techniques, including colored products, chemiluminescence, or autoradiography. Often, the antibodies are labeled with enzymes. When a chemiluminescent substrate is exposed to the enzyme it allows detection. Using western blotting techniques allows not only detection but also quantitative analysis. Analogous methods to western blotting can be used to directly stain specific proteins in live cells or tissue sections. However, these *immunostaining* methods, such as FISH, are used more often in cell biology research.

Eastern Blotting

The Eastern blotting technique is used to detect post-translational modification of proteins. Proteins blotted on to the PVDF or nitrocellulose membrane are probed for modifications using specific substrates.

Microarrays

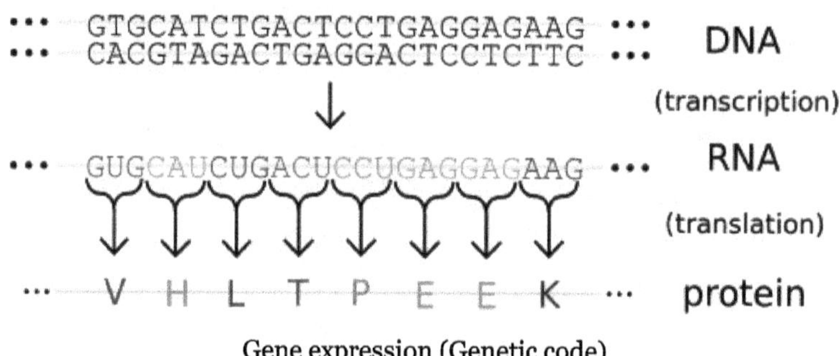

Gene expression (Genetic code)

A DNA microarray is a collection of spots attached to a solid support such as a microscope slide where each spot contains one or more single-stranded DNA oligonucleotide fragment. Arrays make it possible to put down large quantities of very small (100 micrometre diameter) spots on a single slide. Each spot has a DNA fragment molecule that is complementary to a single DNA sequence (similar to Southern blotting). A variation of this technique allows the gene expression of an organism at a particular stage in development to be qualified (expression profiling). In this technique the RNA in a tissue is isolated and converted to labeled cDNA. This cDNA is then hybridized to the fragments on the array and visualization of the hybridization can be done. Since multiple arrays can be made with exactly the same position of fragments they are particularly useful for comparing the gene expression of two different tissues, such as a healthy and cancerous tissue. Also, one can measure what genes are expressed and how that expression changes with time or with other factors. For instance, the common baker's yeast, *Saccharomyces cerevisiae*, contains about 7000 genes; with a microarray, one can measure qualitatively how each gene is expressed, and how that expression changes, for example, with a change in temperature. There are many different ways to fabricate microarrays; the most common are silicon chips, microscope slides with spots of ~ 100 micrometre diameter, custom arrays, and arrays with larger spots on porous membranes (macroarrays). There can be anywhere from 100 spots to more than 10,000 on a given array.Arrays can also be made with molecules other than DNA. For example, an antibody array can be used to determine what proteins or bacteria are present in a blood sample.

Allele-specific Oligonucleotide

Allele-specific oligonucleotide (ASO) is a technique that allows detection of single base mutations without the need for PCR or gel electrophoresis. Short (20-25 nucleotides in length), labeled probes are exposed to the non-fragmented target DNA. Hybridization occurs with high specificity due to the short length of the probes and even a single base change will hinder hybridization. The target DNA is then washed and the labeled probes that didn't hybridize are removed. The target DNA is then analyzed for the presence of the probe via radioactivity or fluorescence. In this experiment, as in most molecular biology techniques, a control must be used to ensure successful experimentation. The Illumina Methylation Assay is an example of a method that takes advantage of the ASO technique to measure one base pair differences in sequence.

Antiquated Technologies

In molecular biology, procedures and technologies are continually being developed and older technologies abandoned. For example, before the advent of DNA gel electrophoresis (agarose or polyacrylamide), the size of DNA molecules was typically determined by rate sedimentation in sucrose gradients, a slow and labor-intensive technique requiring expensive instrumentation; prior to sucrose gradients, viscometry was used.Aside from their historical interest, it is often worth knowing about older technology, as it is occasionally useful to solve another new problem for which the newer technique is inappropriate.

History

While molecular biology was established in the 1930s, the term was coined by Warren Weaver in 1938. Weaver was the director of Natural Sciences for the Rockefeller Foundation at the time and believed that biology was about to undergo a period of significant change given recent advances in fields such as X-ray crystallography. He therefore channeled significant amounts of (Rockefeller Institute) money into biological fields.

Clinical Significance

Clinical research and medical therapies arising from molecular biology are partly covered under gene therapy. The use of molecular biology or molecular cell biology approaches in medicine is now called molecular medicine. Molecular biology also plays important role in understanding formations, actions, regulations of various parts of cells which can be used efficiently for targeting new drugs, diagnosis of disease, physiology of the Cell.

Molecular Microbiology

Molecular microbiology is the branch of microbiology devoted to the study of the molecular basis of the physiological processes that occur in microorganisms.

Bacteria

Mainly because of their relative simplicity, ease of manipulation and growth in vitro, and importance in medicine, bacteria were instrumental in the development of molecular biology. The complete genome sequence for a large number of bacterial species is now available. A list of sequenced prokaryotic genomes is available. Molecular microbiology techniques are currently being used in the development of new genetically engineered vaccines, in bioremediation, biotechnology, food microbiology, probiotic research, antibacterial development and environmental microbiology.

Many bacteria have become model organisms for molecular studies.

Molecular techniques have had a direct influence on the clinical practice of medical microbiology. In many cases where traditional phenotypic methods of microbial identification and typing are insufficient or time-consuming, molecular techniques can provide rapid and accurate data, potentially improving clinical outcomes. Specific examples include:

- 16s rRNA sequencing to provide bacterial identifications
- Pulsed Field Gel Electrophoresis for strain typing of epidemiologically related organisms.
- Direct detection of genes related to resistance mechanisms, such as mecA gene in Staphylococcus aureus

Signaling

Bacteria possess diverse proteins and RNA that can sense changes to their intracellular and extracellular environment. The signals received by these macromolecules are transmitted to key genes or proteins, which alter their activities to suit the new conditions.

Viruses

Viruses are important pathogens of humans and animals. Their genomes are relatively small. For these reasons they were among the first organisms to be fully sequenced. The complete DNA sequence of the Epstein-Barr virus was completed in 1984. Bluetongue virus (BTV) has been in the forefront of molecular studies for last three decades and now represents one of the best understood viruses at the molecular and structural levels. Other viruses such as Papillomavirus, Coronavirus, Caliciviruses, Paramyxoviruses and Influenza virus have also been extensively studied at the molecular level.

Bacterial viruses, or bacteriophages, are estimated to be the most widely distributed and diverse entities in the biosphere. Bacteriophages, or "phage", have been fundamental in the development of the science of molecular biology and became "model organisms" for probing the basic chemistry of life. The first DNA-genome project to be completed was the phage Φ-X174 in 1977. Φ29 phage, a phage of Bacillus, is a paradigm for the study of several molecular mechanisms of general biological processes, including DNA replication and regulation of transcription.

Gene Therapy

Some viruses are used as vectors for gene therapy. Virus vectors have been developed that mediate stable genetic modification of treated cells by chromosomal integration of the transferred vector genomes. Gammaretroviral and lentiviral vectors, for example, can be utilized in clinical gene therapy aimed at the long-term correction of genetic defects, e.g., in stem and progenitor cells. Gammaretroviral and lentiviral vectors have so far been used in more than 300 clinical trials, addressing treatment options for various diseases.

Fungi

Yeasts and molds are eukaryotic microorganisms classified in the kingdom Fungi.

Technology

Polymerase chain reaction (PCR) is used in microbiology to amplify (replicate many times) a single DNA sequence. If required, the sequence can also be altered in predetermined ways. Quantitative PCR is used for the rapid detection of microorganisms and is currently employed in diagnostic clinical microbiology laboratories, environmental analysis, food microbiology, and many other fields. The closely related technique of quantitative PCR permits the quantitative measurement of DNA or RNA molecules and is used to estimate the densities of the reference pathogens in food, water and environmental samples. Quantitative PCR provides both specificity and quantification of target microorganisms.

Gel electrophoresis is used routinely in microbiology to separate DNA, RNA, or protein molecules using an electric field by virtue of their size, shape or electric charge.

Southern blotting, northern blotting, western blotting and Eastern blotting are molecular techniques for detecting the presence of microbial DNA sequences (Southern), RNA sequences (northern), protein molecules (western) or protein modifications (Eastern).

DNA microarrays are used in microbiology as the modern alternative to the "blotting" techniques. Microarrays permit the exploration of thousands of sequences at one time. This technique is used in molecular microbiology to detect the presence of pathogens in a sample (air, water, organ tissue, etc.). It is also used to determine the genetic differences between two microbial strains.

DNA sequencing and genomics have been used for many decades in molecular microbiology studies. Due to their relatively small size, viral genomes were the first to be completely analysed by DNA sequencing. A huge range of sequence and genomic data is now available for a number of species and strains of microorganisms.

RNA interference (RNAi) was discovered as a cellular gene regulation mechanism in 1998, but several RNAi-based applications for gene silencing have already made it into clinical trials. RNA interference (RNAi) technology has formed the basis of novel tools for biological research and drug discovery.

Biomolecule

A biomolecule or biological molecule is any molecule that is present in living organisms, including large macromolecules such as proteins, carbohydrates, lipids, and nucleic acids, as well as small molecules such as primary metabolites, secondary metabolites, and natural products. A more general name for this class of material is biological materials. Biomolecules are usually endogenous but may also be exogenous. For example, pharmaceutical drugs may be natural products or semisynthetic (biopharmaceuticals) or they may be totally synthetic.

Biology and its subsets of biochemistry and molecular biology study biomolecules and their reactions. Most biomolecules are organic compounds, and just four elements—oxygen, carbon, hydrogen, and nitrogen—make up 96% of the human body's mass. But many other elements, such as the various biometals, are present in small amounts.

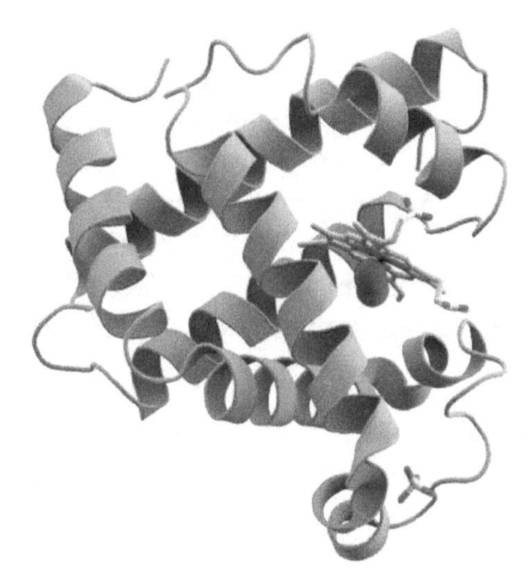

A representation of the 3D structure of myoglobin, showing alpha helices, represented by ribbons. This protein was the first to have its structure solved by X-ray crystallography by Max Perutz and Sir John Cowdery Kendrew in 1958, for which they received a Nobel Prize in Chemistry

Types of Biomolecules

A diverse range of biomolecules exist, including:

- Small molecules:
  - Lipids, fatty acids, glycolipids, sterols, gs
  - Vitamins
  - Hormones, neurotransmitters
  - Metabolites
- Monomers, oligomers and polymers:

Biomonomers	**Bio-oligo**	! Biopolymers	Polymerization **process**	Covalent bond **name between monomers**
Amino acids	Oligopeptides	Polypeptides, proteins (hemoglobin...)	Polycondensation	Peptide bond
Monosaccharides	Oligosaccharides	Polysaccharides (cellulose...)	Polycondensation	Glycosidic bond
Isoprene	Terpenes	Polyterpenes: cis-1,4-polyisoprene natural rubber and trans-1,4-polyisoprene gutta-percha	Polyaddition	
Nucleotides	Oligonucleotides	Polynucleotides, nucleic acids (DNA, RNA)		Phosphodiester bond

Nucleosides and Nucleotides

Nucleosides are molecules formed by attaching a nucleobase to a ribose or deoxyribose ring. Examples of these include cytidine (C), uridine (U), adenosine (A), guanosine (G), thymidine (T) and inosine (I).

Nucleosides can be phosphorylated by specific kinases in the cell, producing nucleotides. Both DNA and RNA are polymers, consisting of long, linear molecules assembled by polymerase enzymes from repeating structural units, or monomers, of mononucleotides. DNA uses the deoxynucleotides C, G, A, and T, while RNA uses the ribonucleotides (which have an extra hydroxyl(OH) group on the pentose ring) C, G, A, and U. Modified bases are fairly common (such as with methyl groups on the base ring), as found in ribosomal RNA or transfer RNAs or for discriminating the new from old strands of DNA after replication.

Each nucleotide is made of an acyclic nitrogenous base, a pentose and one to three phosphate groups. They contain carbon, nitrogen, oxygen, hydrogen and phosphorus. They serve as sources of chemical energy (adenosine triphosphate and guanosine triphosphate), participate in cellular signaling (cyclic guanosine monophosphate and cyclic adenosine monophosphate), and are incorporated into important cofactors of enzymatic reactions (coenzyme A, flavin adenine dinucleotide, flavin mononucleotide, and nicotinamide adenine dinucleotide phosphate).

DNA and RNA Structure

DNA structure is dominated by the well-known double helix formed by Watson-Crick base-pairing of C with G and A with T. This is known as B-form DNA, and is overwhelmingly the most favorable and common state of DNA; its highly specific and stable base-pairing is the basis of reliable genetic information storage. DNA can sometimes occur as single strands (often needing to be stabilized by single-strand binding proteins) or as A-form or Z-form helices, and occasionally in more complex 3D structures such as the crossover at Holliday junctions during DNA replication.

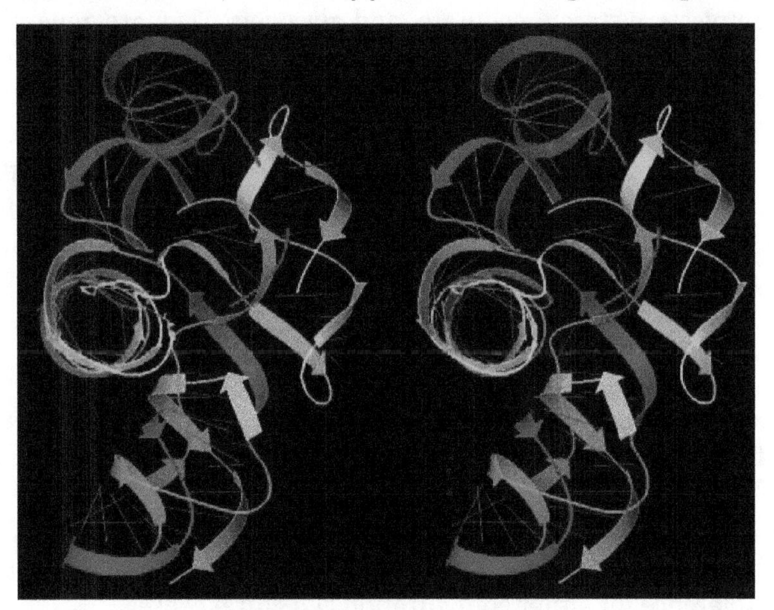

Stereo 3D image of a group I intron ribozyme (PDB file 1Y0Q); gray lines show base pairs; ribbon arrows show double-helix regions, blue to red from 5' to 3' end; white ribbon is an RNA product.

RNA, in contrast, forms large and complex 3D tertiary structures reminiscent of proteins, as well as the loose single strands with locally folded regions that constitute messenger RNA molecules.

Those RNA structures contain many stretches of A-form double helix, connected into definite 3D arrangements by single-stranded loops, bulges, and junctions. Examples are tRNA, ribosomes, ribozymes, and riboswitches. These complex structures are facilitated by the fact that RNA backbone has less local flexibility than DNA but a large set of distinct conformations, apparently because of both positive and negative interactions of the extra OH on the ribose. Structured RNA molecules can do highly specific binding of other molecules and can themselves be recognized specifically; in addition, they can perform enzymatic catalysis (when they are known as "ribozymes", as initially discovered by Tom Cech and colleagues.

Saccharides

Monosaccharides are the simplest form of carbohydrates with only one simple sugar. They essentially contain an aldehyde or ketone group in their structure. The presence of an aldehyde group in a monosaccharide is indicated by the prefix *aldo-*. Similarly, a ketone group is denoted by the prefix *keto-*. Examples of monosaccharides are the hexoses glucose, fructose, and galactose and pentoses, ribose, and deoxyribose Consumed fructose and glucose have different rates of gastric emptying, are differentially absorbed and have different metabolic fates, providing multiple opportunities for 2 different saccharides to differentially affect food intake. Most saccharides eventually provide fuel for cellular respiration.

Disaccharides are formed when two monosaccharides, or two single simple sugars, form a bond with removal of water. They can be hydrolyzed to yield their saccharin building blocks by boiling with dilute acid or reacting them with appropriate enzymes. Examples of disaccharides include sucrose, maltose, and lactose.

Polysaccharides are polymerized monosaccharides, or complex carbohydrates. They have multiple simple sugars. Examples are starch, cellulose, and glycogen. They are generally large and often have a complex branched connectivity. Because of their size, polysaccharides are not water-soluble, but their many hydroxy groups become hydrated individually when exposed to water, and some polysaccharides form thick colloidal dispersions when heated in water. Shorter polysaccharides, with 3 - 10 monomers, are called oligosaccharides. A fluorescent indicator-displacement molecular imprinting sensor was developed for discriminating saccharides. It successfully discriminated three brands of orange juice beverage. The change in fluorescence intensity of the sensing films resulting is directly related to the saccharide concentration.

Lignin

Lignin is a complex polyphenolic macromolecule composed mainly of beta-O4-aryl linkages. After cellulose, lignin is the second most abundant biopolymer and is one of the primary structural components of most plants. It contains subunits derived from *p*-coumaryl alcohol, coniferyl alcohol, and sinapyl alcohol and is unusual among biomolecules in that it is racemic. The lack of optical activity is due to the polymerization of lignin which occurs via free radical coupling reactions in which there is no preference for either configuration at a chiral center.

Lipids

Lipids (oleaginous) are chiefly fatty acid esters, and are the basic building blocks of biological membranes. Another biological role is energy storage (e.g., triglycerides). Most lipids consist of a polar or hydrophilic head (typically glycerol) and one to three nonpolar or hydrophobic fatty acid tails, and therefore they are amphiphilic. Fatty acids consist of unbranched chains of carbon atoms that are connected by single bonds alone (saturated fatty acids) or by both single and double bonds (unsaturated fatty acids). The chains are usually 14-24 carbon groups long, but it is always an even number.

For lipids present in biological membranes, the hydrophilic head is from one of three classes:

- Glycolipids, whose heads contain an oligosaccharide with 1-15 saccharide residues.
- Phospholipids, whose heads contain a positively charged group that is linked to the tail by a negatively charged phosphate group.
- Sterols, whose heads contain a planar steroid ring, for example, cholesterol.

Other lipids include prostaglandins and leukotrienes which are both 20-carbon fatty acyl units synthesized from arachidonic acid. They are also known as fatty acids

Amino Acids

Amino acids contain both amino and carboxylic acid functional groups. (In biochemistry, the term amino acid is used when referring to those amino acids in which the amino and carboxylate functionalities are attached to the same carbon, plus proline which is not actually an amino acid).

Modified amino acids are sometimes observed in proteins; this is usually the result of enzymatic modification after translation (protein synthesis). For example, phosphorylation of serine by kinases and dephosphorylation by phosphatases is an important control mechanism in the cell cycle. Only two amino acids other than the standard twenty are known to be incorporated into proteins during translation, in certain organisms:

- Selenocysteine is incorporated into some proteins at a UGA codon, which is normally a stop codon.
- Pyrrolysine is incorporated into some proteins at a UAG codon. For instance, in some methanogens in enzymes that are used to produce methane.

Besides those used in protein synthesis, other biologically important amino acids include carnitine (used in lipid transport within a cell), ornithine, GABA and taurine.

Protein Structure

The particular series of amino acids that form a protein is known as that protein's primary structure. This sequence is determined by the genetic makeup of the individual. It specifies the order of side-chain groups along the linear polypeptide "backbone".

Proteins have two types of well-classified, frequently occurring elements of local structure defined by a particular pattern of hydrogen bonds along the backbone: alpha helix and beta sheet. Their number and arrangement is called the secondary structure of the protein. Alpha helices are regular spirals stabilized by hydrogen bonds between the backbone CO group (carbonyl) of one amino acid residue and the backbone NH group (amide) of the i+4 residue. The spiral has about 3.6 amino acids per turn, and the amino acid side chains stick out from the cylinder of the helix. Beta pleated sheets are formed by backbone hydrogen bonds between individual beta strands each of which is in an "extended", or fully stretched-out, conformation. The strands may lie parallel or antiparallel to each other, and the side-chain direction alternates above and below the sheet. Hemoglobin contains only helices, natural silk is formed of beta pleated sheets, and many enzymes have a pattern of alternating helices and beta-strands. The secondary-structure elements are connected by "loop" or "coil" regions of non-repetitive conformation, which are sometimes quite mobile or disordered but usually adopt a well-defined, stable arrangement.

The overall, compact, 3D structure of a protein is termed its tertiary structure or its "fold". It is formed as result of various attractive forces like hydrogen bonding, disulfide bridges, hydrophobic interactions, hydrophilic interactions, van der Waals force etc.

When two or more polypeptide chains (either of identical or of different sequence) cluster to form a protein, quaternary structure of protein is formed. Quaternary structure is an attribute of polymeric (same-sequence chains) or heteromeric (different-sequence chains) proteins like hemoglobin, which consists of two "alpha" and two "beta" polypeptide chains.

Apoenzymes

An apoenzyme (or, generally, an apoprotein) is the protein without any small-molecule cofactors, substrates, or inhibitors bound. It is often important as an inactive storage, transport, or secretory form of a protein. This is required, for instance, to protect the secretory cell from the activity of that protein. Apoenzymes becomes active enzymes on addition of a cofactor. Cofactors can be either inorganic (e.g., metal ions and iron-sulfur clusters) or organic compounds, (e.g., flavin and heme). Organic cofactors can be either prosthetic groups, which are tightly bound to an enzyme, or coenzymes, which are released from the enzyme's active site during the reaction.

Isoenzymes

Isoenzymes, or isozymes, are multiple forms of an enzyme, with slightly different protein sequence and closely similar but usually not identical functions. They are either products of different genes, or else different products of alternative splicing. They may either be produced in different organs or cell types to perform the same function, or several isoenzymes may be produced in the same cell type under differential regulation to suit the needs of changing development or environment. LDH (lactate dehydrogenase) has multiple isozymes, while fetal hemoglobin is an example of a developmentally regulated isoform of a non-enzymatic protein. The relative levels of isoenzymes in blood can be used to diagnose problems in the organ of secretion.

References

- Fratamico PM and Bayles DO (editor). (2005). Foodborne Pathogens: Microbiology and Molecular Biology. Caister Academic Press. ISBN 978-1-904455-00-4.
- Miller, AA; Miller, PF (editor) (2011). Emerging Trends in Antibacterial Discovery: Answering the Call to Arms. Caister Academic Press. ISBN 978-1-904455-89-9.
- Sen, K; Ashbolt, NK (editor) (2010). Environmental Microbiology: Current Technology and Water Applications. Caister Academic Press. ISBN 978-1-904455-70-7.

- Mettenleiter TC and Sobrino F (editors). (2008). Animal Viruses: Molecular Biology. Caister Academic Press. ISBN 978-1-904455-22-6.
- Roy P (2008). "Molecular Dissection of Bluetongue Virus". Animal Viruses: Molecular Biology. Caister Academic Press. ISBN 978-1-904455-22-6.
- Campo MS (editor). (2006). Papillomavirus Research: From Natural History To Vaccines and Beyond. Caister Academic Press. ISBN 978-1-904455-04-2.
- Hansman, GS (editor) (2010). Caliciviruses: Molecular and Cellular Virology. Caister Academic Press. ISBN 978-1-904455-63-9.
- Mc Grath S and van Sinderen D (editors). (2007). Bacteriophage: Genetics and Molecular Biology. Caister Academic Press. ISBN 978-1-904455-14-1.
- Kurth, R; Bannert, N (editors) (2010). Retroviruses: Molecular Biology, Genomics and Pathogenesis. Caister Academic Press. ISBN 978-1-904455-55-4.
- Desport, M (editors) (2010). Lentiviruses and Macrophages: Molecular and Cellular Interactions. Caister Academic Press. ISBN 978-1-904455-60-8.
- Kennedy, S; Oswald, N (editor) (2011). PCR Troubleshooting and Optimization: The Essential Guide. Caister Academic Press. ISBN 978-1-904455-72-1.
- Mackay IM (editor). (2007). Real-Time PCR in Microbiology: From Diagnosis to Characterization. Caister Academic Press. ISBN 978-1-904455-18-9.
- Herold, KE; Rasooly, A (editor) (2009). Lab-on-a-Chip Technology: Biomolecular Separation and Analysis. Caister Academic Press. ISBN 978-1-904455-47-9.
- Morris, KV (editor) (2008). RNA and the Regulation of Gene Expression: A Hidden Layer of Complexity. Caister Academic Press. ISBN 978-1-904455-25-7.
- Slabaugh, Michael R. & Seager, Spencer L. (2007). Organic and Biochemistry for Today (6th ed.). Pacific Grove: Brooks Cole. ISBN 0-495-11280-1.
- Alberts B, Johnson A, Lewis J, Raff M, Roberts K & Wlater P (2002). Molecular biology of the cell (4th ed.). New York: Garland Science. pp. 120–1. ISBN 0-8153-3218-1.

2

Processes of Molecular Biology

The processes involved in molecular biology are DNA replication, transcription, gene expression and translation. The biological process of producing two identical replicas of DNA from one original DNA molecule is referred to as DNA replication. This chapter serves as a source to understand the major categories related to molecular biology.

DNA Replication

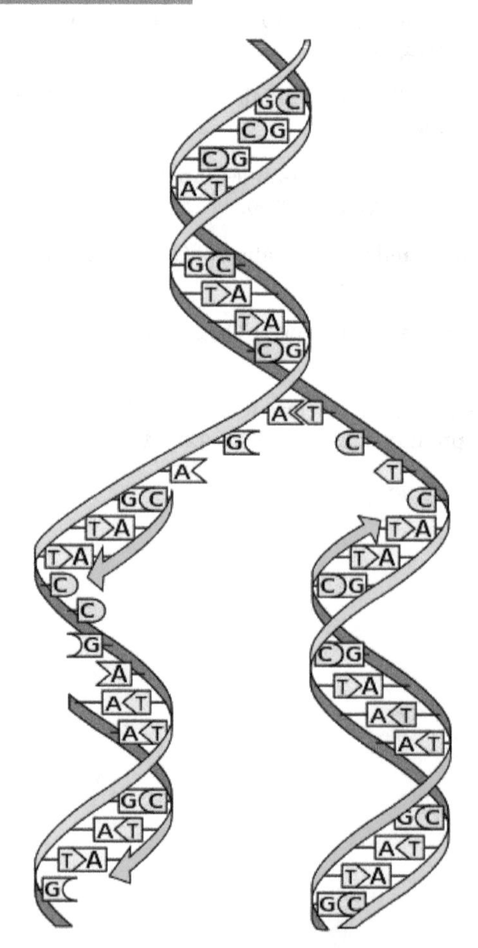

DNA replication. The double helix is unwound and each strand acts as a template for the next strand. Bases are matched to synthesize the new partner strands.

In molecular biology, DNA replication is the biological process of producing two identical replicas of DNA from one original DNA molecule. This process occurs in all living organisms and is the basis for biological inheritance. DNA is made up of a double helix of two complementary strands.

During replication, these strands are separated. Each strand of the original DNA molecule then serves as a template for the production of its counterpart, a process referred to as semiconservative replication. Cellular proofreading and error-checking mechanisms ensure near perfect fidelity for DNA replication.

In a cell, DNA replication begins at specific locations, or origins of replication, in the genome. Unwinding of DNA at the origin and synthesis of new strands results in replication forks growing bi-directionally from the origin. A number of proteins are associated with the replication fork to help in the initiation and continuation of DNA synthesis. Most prominently, DNA polymerase synthesizes the new DNA by adding complementary nucleotides to the template strand. DNA replication occurs during the S-stage of interphase.

DNA replication can also be performed *in vitro* (artificially, outside a cell). DNA polymerases isolated from cells and artificial DNA primers can be used to initiate DNA synthesis at known sequences in a template DNA molecule. The polymerase chain reaction (PCR), a common laboratory technique, cyclically applies such artificial synthesis to amplify a specific target DNA fragment from a pool of DNA.

DNA Structures

DNA usually exists as a double-stranded structure, with both strands coiled together to form the characteristic double-helix. Each single strand of DNA is a chain of four types of nucleotides. Nucleotides in DNA contain a deoxyribose sugar, a phosphate, and a nucleobase. The four types of nucleotide correspond to the four nucleobases adenine, cytosine, guanine, and thymine, commonly abbreviated as A,C, G and T. Adenine and guanine are purine bases, while cytosine and thymine are pyrimidines. These nucleotides form phosphodiester bonds, creating the phosphate-deoxyribose backbone of the DNA double helix with the nucleobases pointing inward. Nucleotides (bases) are matched between strands through hydrogen bonds to form base pairs. Adenine pairs with thymine (two hydrogen bonds), and guanine pairs with cytosine (stronger: three hydrogen bonds).

DNA strands have a directionality, and the different ends of a single strand are called the "3' (three-prime) end" and the "5' (five-prime) end". By convention, if the base sequence of a single strand of DNA is given, the left end of the sequence is the 5' end, while the right end of the sequence is the 3' end. The strands of the double helix are anti-parallel with one being 5' to 3', and the opposite strand 3' to 5'. These terms refer to the carbon atom in deoxyribose to which the next phosphate in the chain attaches. Directionality has consequences in DNA synthesis, because DNA polymerase can synthesize DNA in only one direction by adding nucleotides to the 3' end of a DNA strand.

The pairing of complementary bases in DNA through hydrogen bonding means that the information contained within each strand is redundant. The nucleotides on a single strand can therefore be used to reconstruct nucleotides on a newly synthesized partner strand.

DNA Polymerase

DNA polymerases are a family of enzymes that carry out all forms of DNA replication. DNA polymerases in general cannot initiate synthesis of new strands, but can only extend an existing DNA or RNA strand paired with a template strand. To begin synthesis, a short fragment of RNA, called a primer, must be created and paired with the template DNA strand.

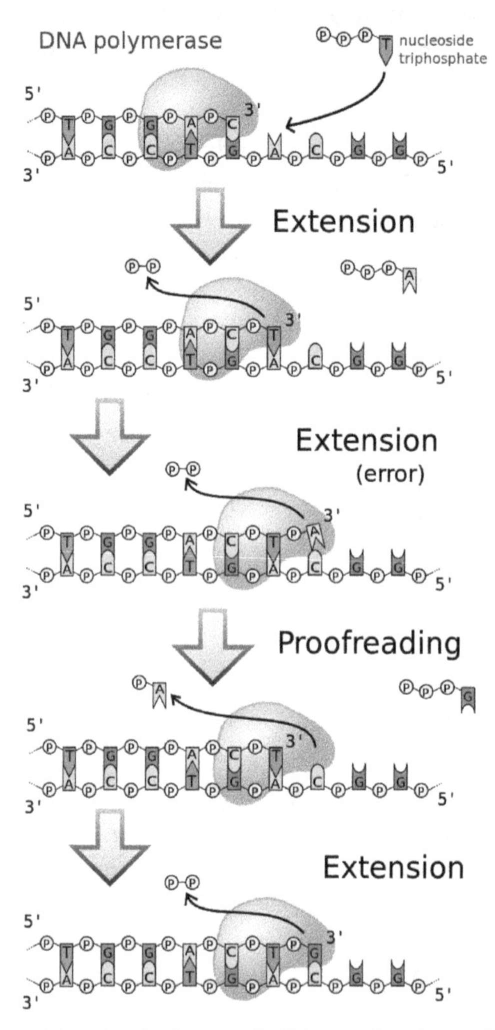

DNA polymerases adds nucleotides to the 3' end of a strand of DNA. If a mismatch is accidentally incorporated, the polymerase is inhibited from further extension. Proofreading removes the mismatched nucleotide and extension continues.

DNA polymerase adds a new strand of DNA by extending the 3' end of an existing nucleotide chain, adding new nucleotides matched to the template strand one at a time via the creation of phosphodiester bonds. The energy for this process of DNA polymerization comes from hydrolysis of the high-energy phosphate (phosphoanhydride) bonds between the three phosphates attached to each unincorporated base. (Free bases with their attached phosphate groups are called nucleotides; in particular, bases with three attached phosphate groups are called nucleoside triphosphates.) When a nucleotide is being added to a growing DNA strand, the formation of a phosphodiester bond between the proximal phosphate of the nucleotide to the growing chain is accompanied by hydrolysis of a high-energy phosphate bond with release of the two distal phosphates as a pyrophosphate. Enzymatic hydrolysis of the resulting pyrophosphate into inorganic phosphate consumes a second high-energy phosphate bond and renders the reaction effectively irreversible.

In general, DNA polymerases are highly accurate, with an intrinsic error rate of less than one mistake for every 10^7 nucleotides added. In addition, some DNA polymerases also have proofreading ability; they can remove nucleotides from the end of a growing strand in order to correct mismatched bases. Finally, post-replication mismatch repair mechanisms monitor the DNA for errors, being capable of distinguishing mismatches in the newly synthesized DNA strand from the original strand sequence. Together, these three discrimination steps enable replication fidelity of less than one mistake for every 10^9 nucleotides added.

The rate of DNA replication in a living cell was first measured as the rate of phage T4 DNA elongation in phage-infected E. coli. During the period of exponential DNA increase at 37 °C, the rate was 749 nucleotides per second. The mutation rate per base pair per replication during phage T4 DNA synthesis is 1.7 per 10^8.

Replication Process

DNA replication, like all biological polymerization processes, proceeds in three enzymatically catalyzed and coordinated steps: initiation, elongation and termination.

Initiation

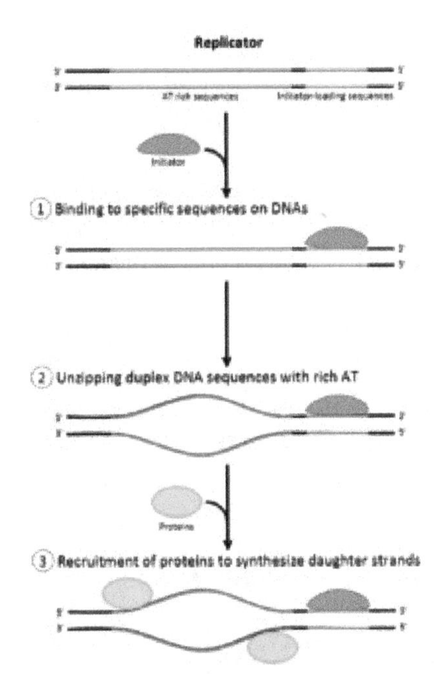

Role of initiators for initiation of DNA replication.

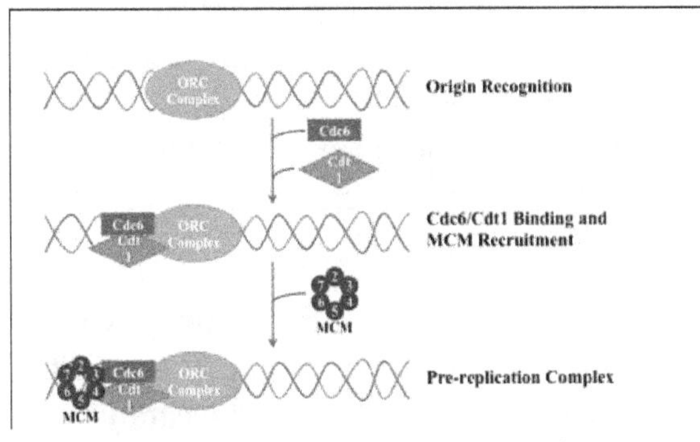

Formation of pre-replication complex.

For a cell to divide, it must first replicate its DNA. This process is initiated at particular points in the DNA, known as "origins", which are targeted by initiator proteins. In *E. coli* this protein is DnaA; in yeast, this is the origin recognition complex. Sequences used by initiator proteins tend to be "AT-rich" (rich in adenine and thymine bases), because A-T base pairs have two hydrogen bonds (rather than the three formed in a C-G pair) which are easier to unzip. Once the origin has been located, these initiators recruit other proteins and form the pre-replication complex, which unzips the double-stranded DNA.

Elongation

DNA polymerase has 5'-3' activity. All known DNA replication systems require a free 3' hydroxyl group before synthesis can be initiated (Important note: the DNA template is read in 3' to 5' direction whereas a new strand is synthesized in the 5' to 3' direction—this is often confused). Four distinct mechanisms for initiation of synthesis are recognized:

1. All cellular life forms and many DNA viruses, phages and plasmids use a primase to synthesize a short RNA primer with a free 3' OH group which is subsequently elongated by a DNA polymerase.

2. The retroelements (including retroviruses) employ a transfer RNA that primes DNA replication by providing a free 3′ OH that is used for elongation by the reverse transcriptase.

3. In the adenoviruses and the φ29 family of bacteriophages, the 3' OH group is provided by the side chain of an amino acid of the genome attached protein (the terminal protein) to which nucleotides are added by the DNA polymerase to form a new strand.

4. In the single stranded DNA viruses — a group that includes the circoviruses, the geminiviruses, the parvoviruses and others — and also the many phages and plasmids that use the rolling circle replication (RCR) mechanism, the RCR endonuclease creates a nick in the genome strand (single stranded viruses) or one of the DNA strands (plasmids). The 5′ end of the nicked strand is transferred to a tyrosine residue on the nuclease and the free 3′ OH group is then used by the DNA polymerase to synthesize the new strand.

The first is the best known of these mechanisms and is used by the cellular organisms. In this mechanism, once the two strands are separated, primase adds RNA primers to the template strands. The leading strand receives one RNA primer while the lagging strand receives several. The leading strand is continuously extended from the primer by a DNA polymerase with high processivity, while the lagging strand is extended discontinuously from each primer forming Okazaki fragments. RNase removes the primer RNA fragments, and a low processivity DNA polymerase distinct from the replicative polymerase enters to fill the gaps. When this is complete, a single nick on the leading strand and several nicks on the lagging strand can be found. Ligase works to fill these nicks in, thus completing the newly replicated DNA molecule.

The primase used in this process differs significantly between bacteria and archaea/eukaryotes. Bacteria use a primase belonging to the DnaG protein superfamily which contains a catalytic domain of the TOPRIM fold type. The TOPRIM fold contains an α/β core with four conserved strands in a Rossmann-like topology. This structure is also found in the catalytic domains of topoisomerase Ia, topoisomerase II, the OLD-family nucleases and DNA repair proteins related to the RecR protein.

The primase used by archaea and eukaryotes, in contrast, contains a highly derived version of the RNA recognition motif (RRM). This primase is structurally similar to many viral RNA-dependent RNA polymerases, reverse transcriptases, cyclic nucleotide generating cyclases and DNA polymerases of the A/B/Y families that are involved in DNA replication and repair. In eukaryotic replication, the primase forms a complex with Pol α.

Multiple DNA polymerases take on different roles in the DNA replication process. In *E. coli*, DNA Pol III is the polymerase enzyme primarily responsible for DNA replication. It assembles into a replication complex at the replication fork that exhibits extremely high processivity, remaining intact for the entire replication cycle. In contrast, DNA Pol I is the enzyme responsible for replacing RNA primers with DNA. DNA Pol I has a 5' to 3' exonuclease activity in addition to its polymerase activity, and uses its exonuclease activity to degrade the RNA primers ahead of it as it extends the DNA strand behind it, in a process called nick translation. Pol I is much less processive than Pol III because its primary function in DNA replication is to create many short DNA regions rather than a few very long regions.

In eukaryotes, the low-processivity enzyme, Pol α, helps to initiate replication. The high-processivity extension enzymes are Pol δ and Pol ε.

As DNA synthesis continues, the original DNA strands continue to unwind on each side of the bubble, forming a replication fork with two prongs. In bacteria, which have a single origin of replication on their circular chromosome, this process creates a "theta structure" (resembling the Greek letter theta: θ). In contrast, eukaryotes have longer linear chromosomes and initiate replication at multiple origins within these.>

Replication Fork

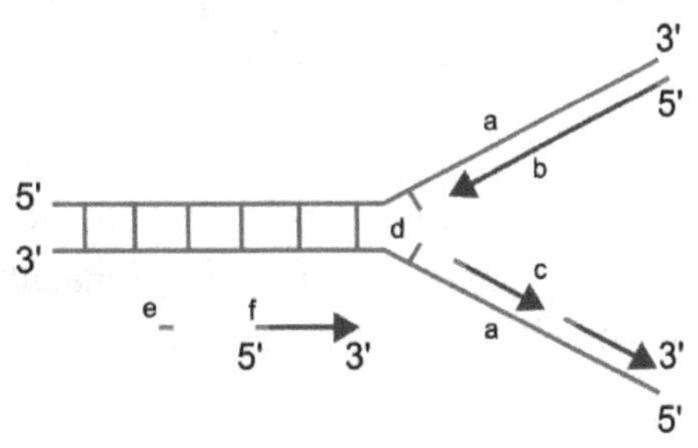

Scheme of the replication fork.
a: template, b: leading strand, c: lagging strand, d: replication fork, e: primer, f: Okazaki fragments

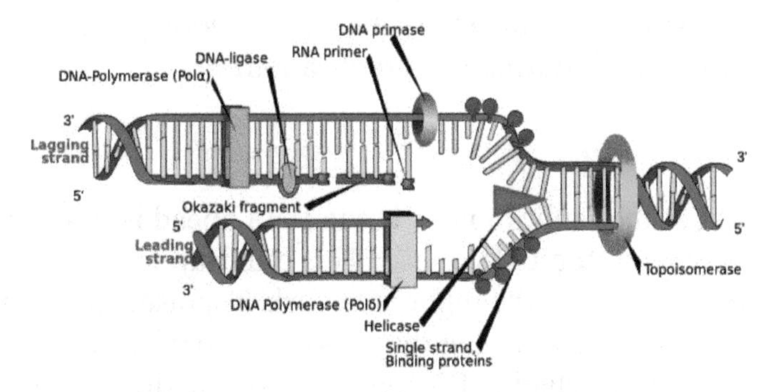

Many enzymes are involved in the DNA replication fork.

The replication fork is a structure that forms within the nucleus during DNA replication. It is created by helicases, which break the hydrogen bonds holding the two DNA strands together. The resulting structure has two branching “prongs”, each one made up of a single strand of DNA. These two strands serve as the template for the leading and lagging strands, which will be created as DNA polymerase matches complementary nucleotides to the templates; the templates may be properly referred to as the leading strand template and the lagging strand template.

DNA is always synthesized in the 5’ to 3’ direction. Since the leading and lagging strand templates are oriented in opposite directions at the replication fork, a major issue is how to achieve synthesis of nascent (new) lagging strand DNA, whose direction of synthesis is opposite to the direction of the growing replication fork.

Leading Strand

The leading strand is the strand of nascent DNA which is being synthesized in the same direction as the growing replication fork. A polymerase “reads” the leading strand *template* and adds complementary nucleotides to the nascent leading strand on a continuous basis.

The polymerase involved in leading strand synthesis is DNA polymerase III (DNA Pol III) in prokaryotes. In eukaryotes, leading strand synthesis is thought to be conducted by Pol ε, however this view has been recently challenged, suggesting a role for Pol δ.

Lagging Strand

The lagging strand is the strand of nascent DNA whose direction of synthesis is opposite to the direction of the growing replication fork. Because of its orientation, replication of the lagging strand is more complicated as compared to that of the leading strand.

The lagging strand is synthesized in short, separated segments. On the lagging strand *template*, a primase “reads” the template DNA and initiates synthesis of a short complementary RNA primer. A DNA polymerase extends the primed segments, forming Okazaki fragments. The RNA primers are then removed and replaced with DNA, and the fragments of DNA are joined together by DNA ligase.

DNA polymerase III (in prokaryotes) or Pol δ (in eukaryotes) is responsible for extension of the primers added during replication of the lagging strand. Primer removal is performed by DNA polymerase I (in prokaryotes) and Pol δ (in eukaryotes). Eukaryotic primase is intrinsic to Pol α. In eukaryotes, pol ε helps with repair during DNA replication.

Dynamics at the Replication Fork

As helicase unwinds DNA at the replication fork, the DNA ahead is forced to rotate. This process results in a build-up of twists in the DNA ahead. This build-up forms a torsional resistance that would eventually halt the progress of the replication fork. Topoisomerases are enzymes that temporarily break the strands of DNA, relieving the tension caused by unwinding the two strands of the DNA helix; topoisomerases (including DNA gyrase) achieve this by adding negative supercoils to the DNA helix.

Bare single-stranded DNA tends to fold back on itself forming secondary structures; these structures can interfere with the movement of DNA polymerase. To prevent this, single-strand binding proteins bind to the DNA until a second strand is synthesized, preventing secondary structure formation.

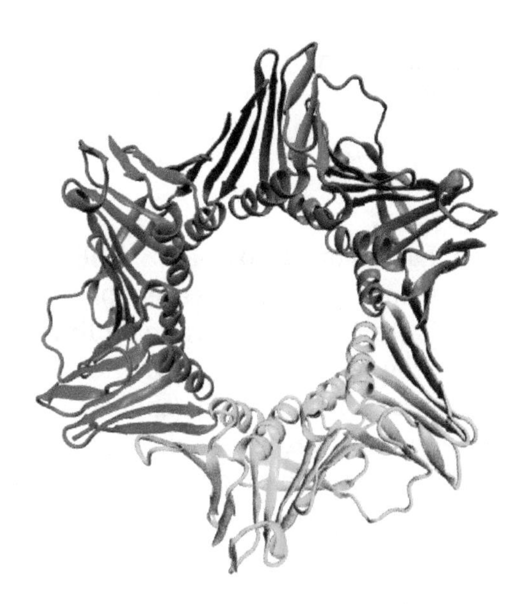

The assembled human DNA clamp, a trimer of the protein PCNA.

Clamp proteins form a sliding clamp around DNA, helping the DNA polymerase maintain contact with its template, thereby assisting with processivity. The inner face of the clamp enables DNA to be threaded through it. Once the polymerase reaches the end of the template or detects double-stranded DNA, the sliding clamp undergoes a conformational change that releases the DNA polymerase. Clamp-loading proteins are used to initially load the clamp, recognizing the junction between template and RNA primers.

DNA Replication Proteins

At the replication fork, many replication enzymes assemble on the DNA into a complex molecular machine called the replisome. The following is a list of major DNA replication enzymes that participate in the replisome:

Enzyme	Function in DNA replication
DNA Helicase	Also known as helix destabilizing enzyme. Unwinds the DNA double helix at the Replication Fork.
DNA Polymerase	Builds a new duplex DNA strand by adding nucleotides in the 5' to 3' direction. Also performs proof-reading and error correction. There exist many different types of DNA Polymerase, each of which perform different functions in different types of cells.
DNA clamp	A protein which prevents elongating DNA polymerases from dissociating from the DNA parent strand.
Single-Strand Binding (SSB) Proteins	Bind to ssDNA and prevent the DNA double helix from re-annealing after DNA helicase unwinds it, thus maintaining the strand separation, and facilitating the synthesis of the nascent strand.
Topoisomerase	Relaxes the DNA from its super-coiled nature.

DNA Gyrase	Relieves strain of unwinding by DNA helicase; this is a specific type of topoisomerase
DNA Ligase	Re-anneals the semi-conservative strands and joins Okazaki Fragments of the lagging strand.
Primase	Provides a starting point of RNA (or DNA) for DNA polymerase to begin synthesis of the new DNA strand.
Telomerase	Lengthens telomeric DNA by adding repetitive nucleotide sequences to the ends of **eukaryotic chromosomes**. This allows germ cells and stem cells to avoid the Hayflick limit on cell division.

Replication Machinery

Replication machineries consist of factors involved in DNA replication and appearing on template ssDNAs. Replication machineries include primosotors are replication enzymes; DNA polymerase, DNA helicases, DNA clamps and DNA topoisomerases, and replication proteins; e.g. single-stranded DNA binding proteins (SSB). In the replication machineries these components coordinate. In most of the bacteria, all of the factors involved in DNA replication are located on replication forks and the complexes stay on the forks during DNA replication. This replication machineries are called replisomes or DNA replicase systems, these terms mean originally a generic term for proteins located on replication forks. In eukaryotic and some bacterial cells the replisomes are not formed.

Since replication machineries do not move relatively to template DNAs such as factories, they are called replication factory. In an alternative figure, DNA factories are similar to projectors and DNAs are like as cinematic films passing constantly into the projectors. In the replication factory model, after both DNA helicases for leading stands and lagging strands are loaded on the template DNAs, the helicases run along the DNAs into each other. The helicases remain associated for the remainder of replication process. Peter Meister et al. observed directly replication sites in budding yeast by monitoring green fluorescent protein(GFP)-tagged DNA polymerases α. They detected DNA replication of pairs of the tagged loci spaced apart symmetrically from a replication origin and found that the distance between the pairs decreased markedly by time. This finding suggests that the mechanism of DNA replication goes with DNA factories. That is, couples of replication factories are loaded on replication origins and the factories associated with each other. Also, template DNAs move into the factories, which bring extrusion of the template ssDNAs and nascent DNAs. Meister's finding is the first direct evidence of replication factory model. Subsequent research has shown that DNA helicases form dimers in many eukaryotic cells and bacterial replication machineries stay in single intranuclear location during DNA synthesis.

The replication factories perform disentanglement of sister chromatids. The disentanglement is essential for distributing the chromatids into daughter cells after DNA replication. Because sister chromatids after DNA replication hold each other by Cohesin rings, there is the only chance for the disentanglement in DNA replication. Fixing of replication machineries as replication factories can improve the success rate of DNA replication. If replication forks move freely in chromosomes, catenation of nuclei is aggravated and impedes mitotic segregation.

Termination

Eukaryotes initiate DNA replication at multiple points in the chromosome, so replication forks meet and terminate at many points in the chromosome; these are not known to be regulated in any particular way. Because eukaryotes have linear chromosomes, DNA replication is unable to reach the very end of the chromosomes, but ends at the telomere region of repetitive DNA close to the ends. This shortens the telomere of the daughter DNA strand. Shortening of the telomeres is a normal process in somatic cells. As a result, cells can only divide a certain number of times before the DNA loss prevents further division. (This is known as the Hayflick limit.) Within the germ cell line, which passes DNA to the next generation, telomerase extends the repetitive sequences of the telomere region to prevent degradation. Telomerase can become mistakenly active in somatic cells, sometimes leading to cancer formation. Increased telomerase activity is one of the hallmarks of cancer.

Termination requires that the progress of the DNA replication fork must stop or be blocked. Termination at a specific locus, when it occurs, involves the interaction between two components: (1) a termination site sequence in the DNA, and (2) a protein which binds to this sequence to physically stop DNA replication. In various bacterial species, this is named the DNA replication terminus site-binding protein, or Ter protein.

Because bacteria have circular chromosomes, termination of replication occurs when the two replication forks meet each other on the opposite end of the parental chromosome. *E. coli* regulates this process through the use of termination sequences that, when bound by the Tus protein, enable only one direction of replication fork to pass through. As a result, the replication forks are constrained to always meet within the termination region of the chromosome.

Regulation

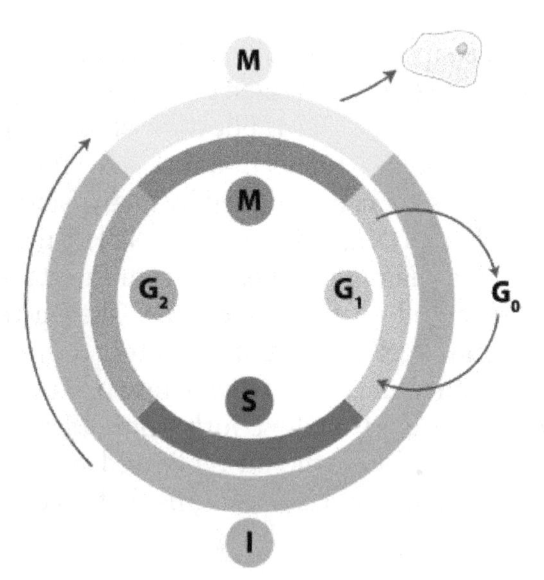

The cell cycle of eukaryotic cells.

Eukaryotes

Within eukaryotes, DNA replication is controlled within the context of the cell cycle. As the cell grows and divides, it progresses through stages in the cell cycle; DNA replication takes place during the S phase (synthesis phase). The progress of the eukaryotic cell through the cycle is controlled by cell cycle checkpoints. Progression through checkpoints is controlled through complex interactions between various proteins, including cyclins and cyclin-dependent kinases. Unlike bacteria, eukaryotic DNA replicates in the confines of the nucleus.

The G1/S checkpoint (or restriction checkpoint) regulates whether eukaryotic cells enter the process of DNA replication and subsequent division. Cells that do not proceed through this checkpoint remain in the G0 stage and do not replicate their DNA.

Replication of chloroplast and mitochondrial genomes occurs independently of the cell cycle, through the process of D-loop replication.

Replication Focus

In vertebrate cells, replication sites concentrate into positions called replication foci. Replication sites can be detected by immunostaining daughter strands and replication enzymes and monitoring GFP-tagged replication factors. By these methods it is found that replication foci of varying size and positions appear in S phase of cell division and their number per nucleus is far smaller than the number of genomic replication forks.

P. Heun et al.(2001) tracked GFP-tagged replication foci in budding yeast cells and revealed that replication origins move constantly in G1 and S phase and the dynamics decreased significantly in S phase. Traditionally, replication sites were fixed on spatial structure of chromosomes by nuclear matrix or lamins. The Heun's results denied the traditional concepts, budding yeasts don't have lamins, and support that replication origins self-assemble and form replication foci.

By firing of replication origins, controlled spatially and temporally, the formation of replication foci is regulated. D. A. Jackson et al.(1998) revealed that neighboring origins fire simultaneously in mammalian cells. Spatial juxtaposition of replication sites brings clustering of replication forks. The clustering do rescue of stalled replication forks and favors normal progress of replication forks. Progress of replication forks is inhibited by many factors; collision with proteins or with complexes binding strongly on DNA, deficiency of dNTPs, nicks on template DNAs and so on. If replication forks stall and remaining sequences from the stalled forks are not replicated, daughter strands have nick obtained un-replicated sites. The un-replicated sites on one parent's strand hold the other strand together but not daughter strands. Therefore, resulting sister chromatids cannot separate together and cannot divide into 2 daughter cells. When neighboring origins fire and a fork from one origin is stalled, fork from other origin access on an opposite direction of the stalled fork and duplicate the un-replicated sites. As other mechanism of the rescue there is application of dormant replication origins that excess origins don't fire in normal DNA replication.

Bacteria

Most bacteria do not go through a well-defined cell cycle but instead continuously copy their DNA; during rapid growth, this can result in the concurrent occurrence of multiple rounds of replication. In *E. coli*, the best-characterized bacteria, DNA replication is regulated through several mechanisms, including: the hemimethylation and sequestering of the origin sequence, the ratio of adenosine triphosphate (ATP) to adenosine diphosphate (ADP), and the levels of protein DnaA. All these control the binding of initiator proteins to the origin sequences.

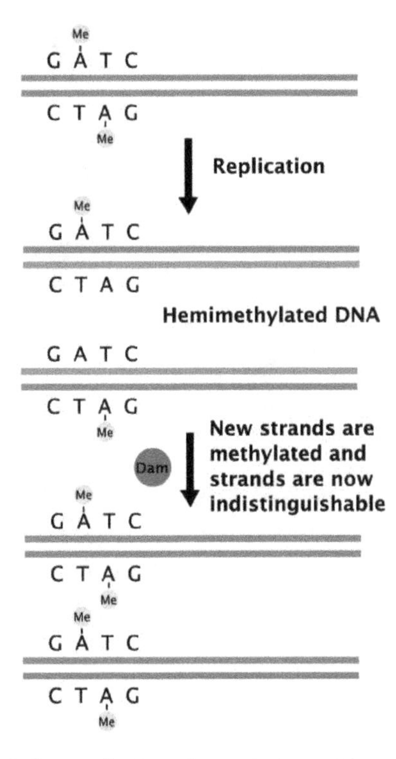

Dam methylates adenine of GATC sites after replication.

Because *E. coli* methylates GATC DNA sequences, DNA synthesis results in hemimethylated sequences. This hemimethylated DNA is recognized by the protein SeqA, which binds and sequesters the origin sequence; in addition, DnaA (required for initiation of replication) binds less well to hemimethylated DNA. As a result, newly replicated origins are prevented from immediately initiating another round of DNA replication.

ATP builds up when the cell is in a rich medium, triggering DNA replication once the cell has reached a specific size. ATP competes with ADP to bind to DnaA, and the DnaA-ATP complex is able to initiate replication. A certain number of DnaA proteins are also required for DNA replication – each time the origin is copied, the number of binding sites for DnaA doubles, requiring the synthesis of more DnaA to enable another initiation of replication.

Polymerase Chain Reaction

Researchers commonly replicate DNA *in vitro* using the polymerase chain reaction (PCR). PCR uses a pair of primers to span a target region in template DNA, and then polymerizes partner strands in each direction from these primers using a thermostable DNA polymerase. Repeating this process through multiple cycles amplifies the targeted DNA region. At the start of each cycle, the mixture of template and primers is heated, separating the newly synthesized molecule and template. Then, as the mixture cools, both of these become templates for annealing of new primers, and the polymerase extends from these. As a result, the number of copies of the target region doubles each round, increasing exponentially.

Transcription (Genetics)

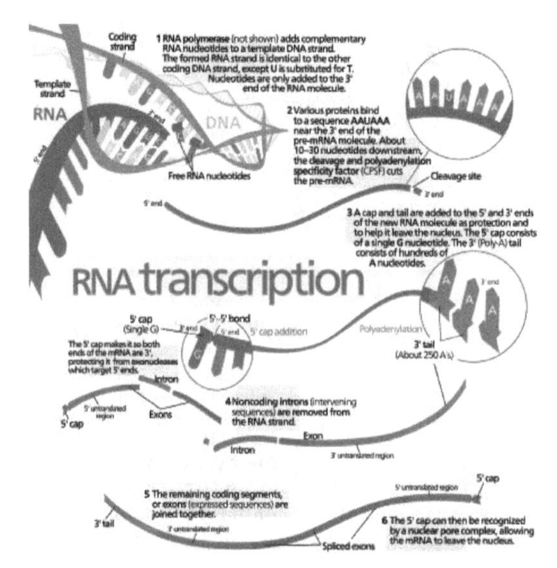

Simplified diagram of mRNA synthesis and processing. Enzymes not shown.

Transcription is the first step of gene expression, in which a particular segment of DNA is copied into RNA (mRNA) by the enzyme RNA polymerase.

Both RNA and DNA are nucleic acids, which use base pairs of nucleotides as a complementary language. The two can be converted back and forth from DNA to RNA by the action of the correct enzymes. During transcription, a DNA sequence is read by an RNA polymerase, which produces a complementary, antiparallel RNA strand called a primary transcript.

Transcription proceeds in the following general steps:

1. RNA polymerase, together with one or more general transcription factor, binds to promoter DNA.
2. RNA polymerase creates a transcription bubble, which separates the two strands of the DNA helix. This is done by breaking the hydrogen bonds between complementary DNA nucleotides.
3. RNA polymerase adds matching RNA nucleotides to the complementary nucleotides of one DNA strand.
4. RNA sugar-phosphate backbone forms with assistance from RNA polymerase to form an RNA strand.
5. Hydrogen bonds of the RNA–DNA helix break, freeing the newly synthesized RNA strand.
6. If the cell has a nucleus, the RNA may be further processed. This may include polyadenylation, capping, and splicing.
7. The RNA may remain in the nucleus or exit to the cytoplasm through the nuclear pore complex.

The stretch of DNA transcribed into an RNA molecule is called a *transcription unit* and encodes at least one gene. If the gene transcribed encodes a protein, messenger RNA (mRNA) will be transcribed; the mRNA will in turn serve as a template for the protein's synthesis through translation. Alternatively, the transcribed gene may encode for either non-coding RNA (such as microRNA), ribosomal RNA (rRNA), transfer RNA (tRNA), or other enzymatic RNA molecules called ribozymes. Overall, RNA helps synthesize, regulate, and process proteins; it therefore plays a fundamental role in performing functions within a cell.

In virology, the term may also be used when referring to mRNA synthesis from an RNA molecule (i.e., RNA replication). For instance, the genome of a negative-sense single-stranded RNA (ssRNA -) virus may be template for a positive-sense single-stranded RNA (ssRNA +). This is because the positive-sense strand contains the information needed to translate the viral proteins for viral replication afterwards. This process is catalyzed by a viral RNA replicase.

Background

A DNA transcription unit encoding for a protein may contain both a *coding sequence*, which will be translated into the protein, and *regulatory sequences*, which direct and regulate the synthesis of that protein. The regulatory sequence before ("upstream" from) the coding sequence is called the five prime untranslated region (5'UTR); the sequence after ("downstream" from) the coding sequence is called the three prime untranslated region (3'UTR).

As opposed to DNA replication, transcription results in an RNA complement that includes the nucleotide uracil (U) in all instances where thymine (T) would have occurred in a DNA complement.

Only one of the two DNA strands serve as a template for transcription. The antisense strand of DNA is read by RNA polymerase from the 3' end to the 5' end during transcription (3' → 5'). The complementary RNA is created in the opposite direction, in the 5' → 3' direction, matching the sequence of the sense strand with the exception of switching uracil for thymine. This directionality is because RNA polymerase can only add nucleotides to the 3' end of the growing mRNA chain. This use of only the 3' → 5' DNA strand eliminates the need for the Okazaki fragments that are seen in DNA replication. This removes the need for an RNA primer to initiate RNA synthesis, as is the case in DNA replication.

The *non*-template sense strand of DNA is called the coding strand, because its sequence is the same as the newly created RNA transcript (except for the substitution of uracil for thymine). This is the strand that is used by convention when presenting a DNA sequence.

Transcription has some proofreading mechanisms, but they are fewer and less effective than the controls for copying DNA; therefore, transcription has a lower copying fidelity than DNA replication.

Major Steps

Transcription is divided into *initiation*, *promoter escape*, *elongation* and *termination*.

Initiation

Transcription begins with the binding of RNA polymerase, together with one or more general transcription factor, to a specific DNA sequence referred to as a "promoter" to form an RNA polymerase-promoter "closed complex" (called a "closed complex" because the promoter DNA is fully double-stranded).

RNA polymerase, assisted by one or more general transcription factors, then unwinds approximately 14 base pairs of DNA to form an RNA polymerase-promoter "open complex" (called an "open complex" because the promoter DNA is partly unwound and single-stranded) that contains an unwound, singe-stranded DNA region of approximately 14 base pairs referred to as the "transcription bubble."

RNA polymerase, assisted by one or more general transcription factors, then selects a transcription start site in the transcription bubble, binds to an initiating NTP and an extending NTP (or a short RNA primer and an extending NTP) complementary to the transcription start site sequence, and catalyzes bond formation to yield an initial RNA product.

In bacteria, RNA polymerase core enzyme consists of five subunits: 2 α subunits, 1 β subunit, 1 β' subunit, and 1 ω subunit. In bacteria, there is one general RNA transcription factor: sigma. RNA polymerase core enzyme binds to the bacterial general transcription factor sigma to form RNA polymerase holoenzyme and then binds to a promoter.

In archaea and eukaryotes, RNA polymerase contains subunits homologous to each of the five RNA polymerase subunits in bacteria and also contains additional subunits. In archaea and eukaryotes, the functions performed by the bacterial general transcription factor sigma are performed by multiple general transcription factors that work together. In archaea, there are three general transcription factors: TBP, TFB, and TFE. In eukaryotes, in RNA polymerase II-dependent transcription, there are six general transcription factors: TFIIA, TFIIB (an ortholog of archaeal TFB), TFIID (a multisubunit factor in which the key subunit, TBP, is an ortholog of archaeal TBP), TFIIE (an ortholog of archaeal TFE), TFIIF, and TFIIH. In archaea and eukaryotes, the RNA polymerase-promoter closed complex usually is referred to as the "preinitiation complex."

Transcription initiation is regulated by additional proteins, known as activators and repressors, and, in some cases, associated coactivators or corepressors, which modulate formation and function of the transcription initiation complex.

Promoter Escape

After the first bond is synthesized, the RNA polymerase must escape the promoter. During this time there is a tendency to release the RNA transcript and produce truncated transcripts. This is called *abortive initiation* and is common for both eukaryotes and prokaryotes. Abortive initiation continues to occur until an RNA product of a threshold length of approximately 10 nucleotides is synthesized, at which point promoter escape occurs and a transcription elongation complex is formed.

Mechanistically, promoter escape occurs through a scrunching mechanism, where the energy built up by DNA scrunching provides the energy needed to break interactions between RNA polymerase holoenzyme and the promoter.

In bacteria, upon and following promoter clearance, the σ factor is released according to a stochastic model.

In eukaryotes, at an RNA polymerase II-dependent promoter, upon promoter clearance, TFIIH phosphorylates serine 5 on the carboxy terminal domain of RNA polymerase II, leading to the recruitment of capping enzyme (CE). The exact mechanism of how CE induces promoter clearance in eukaryotes is not yet known.

Elongation

Simple diagram of transcription elongation

One strand of the DNA, the *template strand* (or noncoding strand), is used as a template for RNA synthesis. As transcription proceeds, RNA polymerase traverses the template strand and uses base pairing complementarity with the DNA template to create an RNA copy. Although RNA polymerase traverses the template strand from 3' → 5', the coding (non-template) strand and newly formed RNA can also be used as reference points, so transcription can be described as occurring 5' → 3'. This produces an RNA molecule from 5' → 3', an exact copy of the coding strand (except that thymines are replaced with uracils, and the nucleotides are composed of a ribose (5-carbon) sugar where DNA has deoxyribose (one fewer oxygen atom) in its sugar-phosphate backbone).

mRNA transcription can involve multiple RNA polymerases on a single DNA template and multiple rounds of transcription (amplification of particular mRNA), so many mRNA molecules can be rapidly produced from a single copy of a gene.

Elongation also involves a proofreading mechanism that can replace incorrectly incorporated bases. In eukaryotes, this may correspond with short pauses during transcription that allow appropriate RNA editing factors to bind. These pauses may be intrinsic to the RNA polymerase or due to chromatin structure.

Termination

Bacteria use two different strategies for transcription termination - *Rho-independent termination* and *Rho-dependent termination.* In Rho-independent transcription termination, also called intrinsic termination, RNA transcription stops when the newly synthesized RNA molecule forms a G-C-rich hairpin loop followed by a run of Us. When the hairpin forms, the mechanical stress breaks the weak rU-dA bonds, now filling the DNA–RNA hybrid. This pulls the poly-U transcript out of the active site of the RNA polymerase, in effect, terminating transcription. In the "Rho-dependent" type of termination, a protein factor called "Rho" destabilizes the interaction between the template and the mRNA, thus releasing the newly synthesized mRNA from the elongation complex.

Transcription termination in eukaryotes is less understood but involves cleavage of the new transcript followed by template-independent addition of adenines at its new 3' end, in a process called polyadenylation.

Inhibitors

Transcription inhibitors can be used as antibiotics against, for example, pathogenic bacteria (antibacterials) and fungi (antifungals). An example of such an antibacterial is rifampicin, which inhibits prokaryotic DNA transcription into mRNA by inhibiting DNA-dependent RNA polymerase by binding its beta-subunit. 8-Hydroxyquinoline is an antifungal transcription inhibitor. The effects of histone methylation may also work to inhibit the action of transcription.

Factories

Active transcription units are clustered in the nucleus, in discrete sites called transcription factories or euchromatin. Such sites can be visualized by allowing engaged polymerases to extend their transcripts in tagged precursors (Br-UTP or Br-U) and immuno-labeling the tagged nascent RNA. Transcription factories can also be localized using fluorescence in situ hybridization or marked by antibodies directed against polymerases. There are ~10,000 factories in the nucleoplasm of a HeLa cell, among which are ~8,000 polymerase II factories and ~2,000 polymerase III factories. Each polymerase II factory contains ~8 polymerases. As most active transcription units are associated with only one polymerase, each factory usually contains ~8 different transcription units. These units might be associated through promoters and/or enhancers, with loops forming a 'cloud' around the factor.

History

A molecule that allows the genetic material to be realized as a protein was first hypothesized by François Jacob and Jacques Monod. Severo Ochoa won a Nobel Prize in Physiology or Medicine in 1959 for developing a process for synthesizing RNA *in vitro* with polynucleotide phosphorylase, which was useful for cracking the genetic code. RNA synthesis by RNA polymerase was established *in vitro* by several laboratories by 1965; however, the RNA synthesized by these enzymes had properties that suggested the existence of an additional factor needed to terminate transcription correctly.

In 1972, Walter Fiers became the first person to actually prove the existence of the terminating enzyme.

Roger D. Kornberg won the 2006 Nobel Prize in Chemistry "for his studies of the molecular basis of eukaryotic transcription".

Measuring and Detecting

Transcription can be measured and detected in a variety of ways:

- G-Less Cassette transcription assay: measures promoter strength
- Run-off Transcription assay: identifies transcription start sites (TSS)
- Nuclear Run-on assay: measures the relative abundance of newly formed transcripts
- RNase protection assay and ChIP-Chip of RNAP: detect active transcription sites

- RT-PCR: measures the absolute abundance of total or nuclear RNA levels, which may however differ from transcription rates
- DNA microarrays: measures the relative abundance of the global total or nuclear RNA levels; however, these may differ from transcription rates
- In situ hybridization: detects the presence of a transcript
- MS2 tagging: by incorporating RNA stem loops, such as MS2, into a gene, these become incorporated into newly synthesized RNA. The stem loops can then be detected using a fusion of GFP and the MS2 coat protein, which has a high affinity, sequence-specific interaction with the MS2 stem loops. The recruitment of GFP to the site of transcription is visualized as a single fluorescent spot. This new approach has revealed that transcription occurs in discontinuous bursts, or pulses. With the notable exception of in situ techniques, most other methods provide cell population averages, and are not capable of detecting this fundamental property of genes.
- Northern blot: the traditional method, and until the advent of RNA-Seq, the most quantitative
- RNA-Seq: applies next-generation sequencing techniques to sequence whole transcriptomes, which allows the measurement of relative abundance of RNA, as well as the detection of additional variations such as fusion genes, post-transcriptional edits and novel splice sites

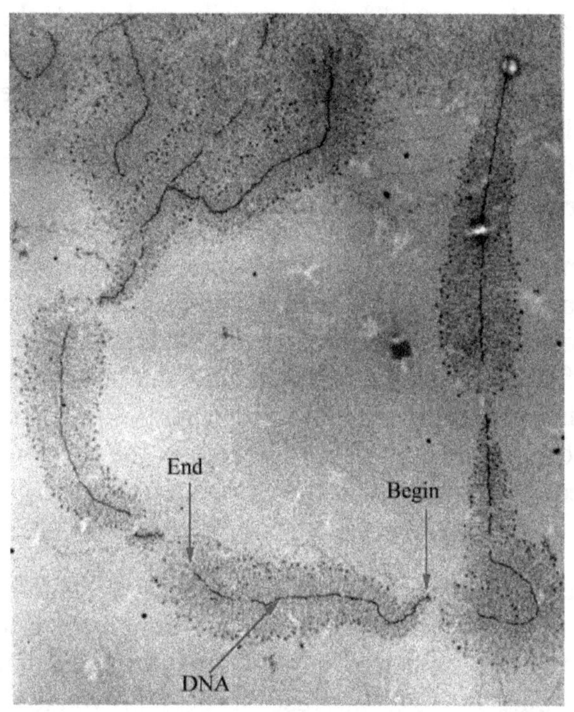

Electron micrograph of transcription of ribosomal RNA. The forming ribosomal RNA strands are visible as branches from the main DNA strand.

Reverse Transcription

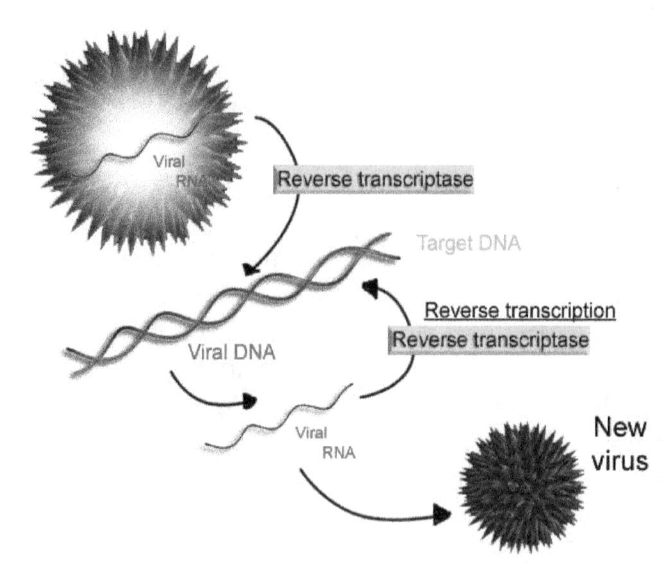

Scheme of reverse transcription

Some viruses (such as HIV, the cause of AIDS), have the ability to transcribe RNA into DNA. HIV has an RNA genome that is *reverse transcribed* into DNA. The resulting DNA can be merged with the DNA genome of the host cell. The main enzyme responsible for synthesis of DNA from an RNA template is called reverse transcriptase.

In the case of HIV, reverse transcriptase is responsible for synthesizing a complementary DNA strand (cDNA) to the viral RNA genome. The enzyme ribonuclease H then digests the RNA strand, and reverse transcriptase synthesises a complementary strand of DNA to form a double helix DNA structure ("cDNA"). The cDNA is integrated into the host cell's genome by the enzyme integrase, which causes the host cell to generate viral proteins that reassemble into new viral particles. In HIV, subsequent to this, the host cell undergoes programmed cell death, or apoptosis of T cells. However, in other retroviruses, the host cell remains intact as the virus buds out of the cell.

Some eukaryotic cells contain an enzyme with reverse transcription activity called telomerase. Telomerase is a reverse transcriptase that lengthens the ends of linear chromosomes. Telomerase carries an RNA template from which it synthesizes a repeating sequence of DNA, or "junk" DNA. This repeated sequence of DNA is called a telomere and can be thought of as a "cap" for a chromosome. It is important because every time a linear chromosome is duplicated, it is shortened. With this "junk" DNA or "cap" at the ends of chromosomes, the shortening eliminates some of the non-essential, repeated sequence rather than the protein-encoding DNA sequence, that is farther away from the chromosome end.

Telomerase is often activated in cancer cells to enable cancer cells to duplicate their genomes indefinitely without losing important protein-coding DNA sequence. Activation of telomerase could be part of the process that allows cancer cells to become *immortal*. The immortalizing factor of cancer via telomere lengthening due to telomerase has been proven to occur in 90% of all carcinogenic tumors *in vivo* with the remaining 10% using an alternative telomere maintenance route called ALT or Alternative Lengthening of Telomeres.

Gene Expression

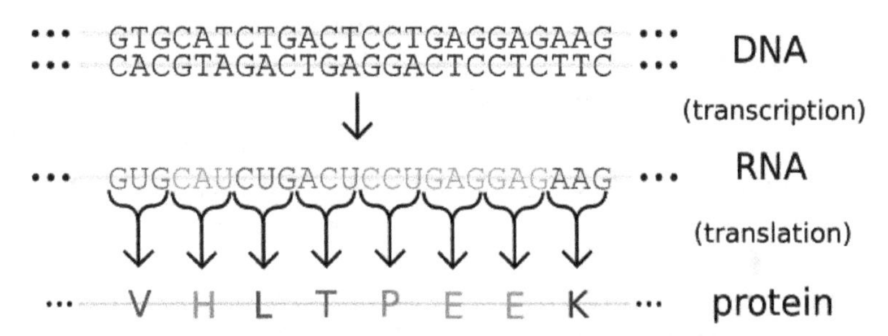

Genes are expressed by being transcribed into RNA, and this transcript may then be translated into protein.

Gene expression is the process by which information from a gene is used in the synthesis of a functional gene product. These products are often proteins, but in non-protein coding genes such as transfer RNA (tRNA) or small nuclear RNA (snRNA) genes, the product is a functional RNA. The process of gene expression is used by all known life—eukaryotes (including multicellular organisms), prokaryotes (bacteria and archaea), and utilized by viruses—to generate the macromolecular machinery for life.

Several steps in the gene expression process may be modulated, including the transcription, RNA splicing, translation, and post-translational modification of a protein. Gene regulation gives the cell control over structure and function, and is the basis for cellular differentiation, morphogenesis and the versatility and adaptability of any organism. Gene regulation may also serve as a substrate for evolutionary change, since control of the timing, location, and amount of gene expression can have a profound effect on the functions (actions) of the gene in a cell or in a multicellular organism.

In genetics, gene expression is the most fundamental level at which the genotype gives rise to the phenotype, i.e. observable trait. The genetic code stored in DNA is "interpreted" by gene expression, and the properties of the expression give rise to the organism's phenotype. Such phenotypes are often expressed by the synthesis of proteins that control the organism's shape, or that act as enzymes catalysing specific metabolic pathways characterising the organism. Regulation of gene expression is thus critical to an organism's development.

Mechanism

Transcription

The process of transcription is carried out by RNA polymerase (RNAP), which uses DNA (black) as a template and produces RNA (blue).

A gene is a stretch of DNA that encodes information. Genomic DNA consists of two antiparallel and reverse complementary strands, each having 5' and 3' ends. With respect to a gene, the two strands may be labeled the "template strand," which serves as a blueprint for the production of an RNA transcript, and the "coding strand," which includes the DNA version of the transcript sequence. The production of RNA copies of the DNA is called transcription, and is performed in the nucleus by RNA polymerase, which adds one RNA nucleotide at a time to a growing RNA strand. This RNA is complementary to the template 3' → 5' DNA strand, which is itself complementary to the coding 5' → 3' DNA strand. Therefore, the resulting 5' → 3' RNA strand is identical to the coding DNA strand with the exception that thymines (T) are replaced with uracils (U) in the RNA. A coding DNA strand reading "ATG" is indirectly transcribed through the non-coding strand as "AUG" in RNA.

Transcription in bacteria is carried out by a single type of RNA polymerase, which needs a DNA sequence called a Pribnow box as well as a sigma factor (σ factor) to start transcription. In eukaryotes, transcription is performed by three types of RNA polymerases, each of which needs a special DNA sequence called the promoter and a set of DNA-binding proteins—transcription factors—to initiate the process. RNA polymerase I is responsible for transcription of ribosomal RNA (rRNA) genes. RNA polymerase II (Pol II) transcribes all protein-coding genes but also some non-coding RNAs (e.g., snRNAs, snoRNAs or long non-coding RNAs). Pol II includes a C-terminal domain (CTD) that is rich in serine residues. When these residues are phosphorylated, the CTD binds to various protein factors that promote transcript maturation and modification. RNA polymerase III transcribes 5S rRNA, transfer RNA (tRNA) genes, and some small non-coding RNAs (e.g., 7SK). Transcription ends when the polymerase encounters a sequence called the terminator.

RNA Processing

While transcription of prokaryotic protein-coding genes creates messenger RNA (mRNA) that is ready for translation into protein, transcription of eukaryotic genes leaves a primary transcript of RNA (pre-mRNA), which first has to undergo a series of modifications to become a mature mRNA.

These include 5' *capping*, which is set of enzymatic reactions that add 7-methylguanosine (m^7G) to the 5' end of pre-mRNA and thus protect the RNA from degradation by exonucleases. The m^7G cap is then bound by cap binding complex heterodimer (CBC20/CBC80), which aids in mRNA export to cytoplasm and also protect the RNA from decapping.

Another modification is 3' *cleavage and polyadenylation*. They occur if polyadenylation signal sequence (5'- AAUAAA-3') is present in pre-mRNA, which is usually between protein-coding sequence and terminator. The pre-mRNA is first cleaved and then a series of ~200 adenines (A) are added to form poly(A) tail, which protects the RNA from degradation. Poly(A) tail is bound by multiple poly(A)-binding proteins (PABP) necessary for mRNA export and translation re-initiation.

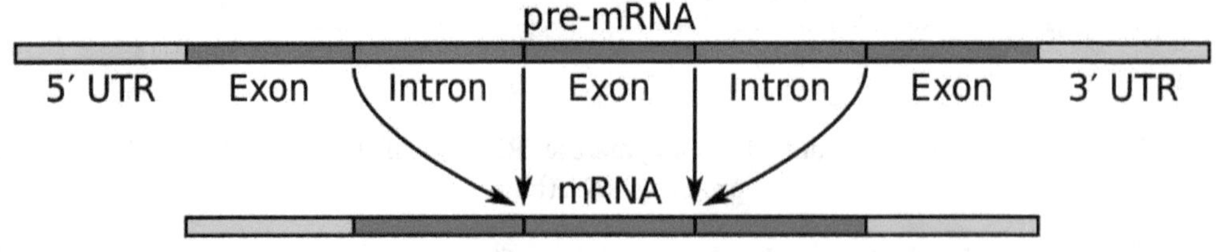

Simple illustration of exons and introns in pre-mRNA and the formation of mature mRNA by splicing. The UTRs are non-coding parts of exons at the ends of the mRNA.

A very important modification of eukaryotic pre-mRNA is *RNA splicing*. The majority of eukaryotic pre-mRNAs consist of alternating segments called exons and introns. During the process of splicing, an RNA-protein catalytical complex known as spliceosome catalyzes two transesterification reactions, which remove an intron and release it in form of lariat structure, and then splice neighbouring exons together. In certain cases, some introns or exons can be either removed or retained in mature mRNA. This so-called alternative splicing creates series of different transcripts originating from a single gene. Because these transcripts can be potentially translated into different proteins, splicing extends the complexity of eukaryotic gene expression.

Extensive RNA processing may be an evolutionary advantage made possible by the nucleus of eukaryotes. In prokaryotes, transcription and translation happen together, whilst in eukaryotes, the nuclear membrane separates the two processes, giving time for RNA processing to occur.

Non-coding RNA Maturation

In most organisms non-coding genes (ncRNA) are transcribed as precursors that undergo further processing. In the case of ribosomal RNAs (rRNA), they are often transcribed as a pre-rRNA that contains one or more rRNAs. The pre-rRNA is cleaved and modified (2′-O-methylation and pseudouridine formation) at specific sites by approximately 150 different small nucleolus-restricted RNA species, called snoRNAs. SnoRNAs associate with proteins, forming snoRNPs. While snoRNA part basepair with the target RNA and thus position the modification at a precise site, the protein part performs the catalytical reaction. In eukaryotes, in particular a snoRNP called RNase, MRP cleaves the 45S pre-rRNA into the 28S, 5.8S, and 18S rRNAs. The rRNA and RNA processing factors form large aggregates called the nucleolus.

In the case of transfer RNA (tRNA), for example, the 5' sequence is removed by RNase P, whereas the 3' end is removed by the tRNase Z enzyme and the non-templated 3' CCA tail is added by a nucleotidyl transferase. In the case of micro RNA (miRNA), miRNAs are first transcribed as primary transcripts or pri-miRNA with a cap and poly-A tail and processed to short, 70-nucleotide stem-loop structures known as pre-miRNA in the cell nucleus by the enzymes Drosha and Pasha. After being exported, it is then processed to mature miRNAs in the cytoplasm by interaction with the endonuclease Dicer, which also initiates the formation of the RNA-induced silencing complex (RISC), composed of the Argonaute protein.

Even snRNAs and snoRNAs themselves undergo series of modification before they become part of functional RNP complex. This is done either in the nucleoplasm or in the specialized compartments called Cajal bodies. Their bases are methylated or pseudouridinilated by a group of small Cajal body-specific RNAs (scaRNAs), which are structurally similar to snoRNAs.

RNA Export

In eukaryotes most mature RNA must be exported to the cytoplasm from the nucleus. While some RNAs function in the nucleus, many RNAs are transported through the nuclear pores and into the cytosol. Notably this includes all RNA types involved in protein synthesis. In some cases RNAs are additionally transported to a specific part of the cytoplasm, such as a synapse; they are then towed by motor proteins that bind through linker proteins to specific sequences (called "zipcodes") on the RNA.

Translation

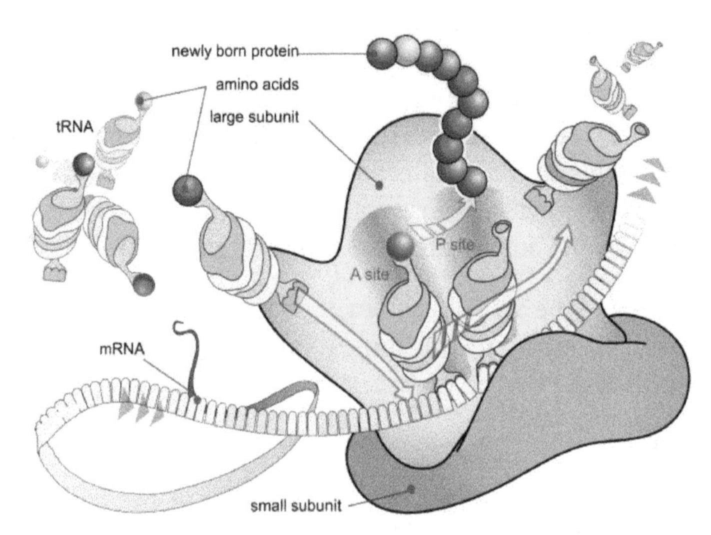

During the translation, tRNA charged with amino acid enters the ribosome and aligns with the correct mRNA triplet. Ribosome then adds amino acid to growing protein chain.

For some RNA (non-coding RNA) the mature RNA is the final gene product. In the case of messenger RNA (mRNA) the RNA is an information carrier coding for the synthesis of one or more proteins. mRNA carrying a single protein sequence (common in eukaryotes) is monocistronic whilst mRNA carrying multiple protein sequences (common in prokaryotes) is known as polycistronic.

Every mRNA consists of three parts: a 5' untranslated region (5'UTR), a protein-coding region or open reading frame (ORF), and a 3' untranslated region (3'UTR). The coding region carries information for protein synthesis encoded by the genetic code to form triplets. Each triplet of nucleotides of the coding region is called a codon and corresponds to a binding site complementary to an anticodon triplet in transfer RNA. Transfer RNAs with the same anticodon sequence always carry an identical type of amino acid. Amino acids are then chained together by the ribosome according to the order of triplets in the coding region. The ribosome helps transfer RNA to bind to messenger RNA and takes the amino acid from each transfer RNA and makes a structure-less protein out of it. Each mRNA molecule is translated into many protein molecules, on average ~2800 in mammals.

In prokaryotes translation generally occurs at the point of transcription (co-transcriptionally), often using a messenger RNA that is still in the process of being created. In eukaryotes translation can occur in a variety of regions of the cell depending on where the protein being written is supposed to be. Major locations are the cytoplasm for soluble cytoplasmic proteins and the membrane of the endoplasmic reticulum for proteins that are for export from the cell or insertion into a cell membrane. Proteins that are supposed to be expressed at the endoplasmic reticulum are recognised part-way through the translation process. This is governed by the signal recognition particle—a protein that binds to the ribosome and directs it to the endoplasmic reticulum when it finds a signal peptide on the growing (nascent) amino acid chain. Translation is the communication of the meaning of a source-language text by means of an equivalent target-language text

Folding

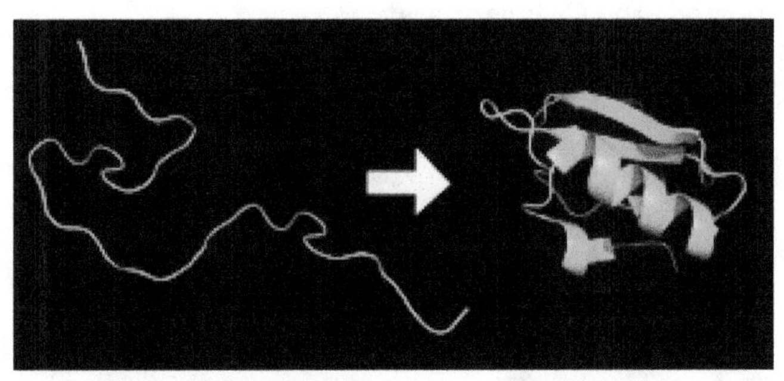

Protein before (left) and after (right) folding.

The polypeptide folds into its characteristic and functional three-dimensional structure from a random coil. Each protein exists as an unfolded polypeptide or random coil when translated from a sequence of mRNA into a linear chain of amino acids. This polypeptide lacks any developed three-dimensional structure (the left hand side of the neighboring figure). Amino acids interact with each other to produce a well-defined three-dimensional structure, the folded protein (the right hand side of the figure) known as the native state. The resulting three-dimensional structure is determined by the amino acid sequence (Anfinsen's dogma).

The correct three-dimensional structure is essential to function, although some parts of functional proteins may remain unfolded Failure to fold into the intended shape usually produces inactive proteins with different properties including toxic prions. Several neurodegenerative and other diseases are believed to result from the accumulation of *misfolded* proteins. Many allergies are caused by the folding of the proteins, for the immune system does not produce antibodies for certain protein structures.

Enzymes called chaperones assist the newly formed protein to attain (fold into) the 3-dimensional structure it needs to function. Similarly, RNA chaperones help RNAs attain their functional shapes. Assisting protein folding is one of the main roles of the endoplasmic reticulum in eukaryotes.

Translocation

Secretory proteins of eukaryotes or prokaryotes must be translocated to enter the secretory pathway. Newly synthesized proteins are directed to the eukaryotic Sec61 or prokaryotic SecYEG translocation channel by signal peptides. The efficiency of protein secretion in eukaryotes is very dependent on the signal peptide which has been used.

Protein Transport

Many proteins are destined for other parts of the cell than the cytosol and a wide range of signalling sequences or (signal peptides) are used to direct proteins to where they are supposed to be. In prokaryotes this is normally a simple process due to limited compartmentalisation of the cell. However, in eukaryotes there is a great variety of different targeting processes to ensure the protein arrives at the correct organelle.

Not all proteins remain within the cell and many are exported, for example, digestive enzymes, hormones and extracellular matrix proteins. In eukaryotes the export pathway is well developed and the main mechanism for the export of these proteins is translocation to the endoplasmic reticulum, followed by transport via the Golgi apparatus.

Regulation of Gene Expression

The patchy colours of a tortoiseshell cat are the result of different levels of expression of pigmentation genes in different areas of the skin.

Regulation of gene expression refers to the control of the amount and timing of appearance of the functional product of a gene. Control of expression is vital to allow a cell to produce the gene products it needs when it needs them; in turn, this gives cells the flexibility to adapt to a variable environment, external signals, damage to the cell, etc. Some simple examples of where gene expression is important are:

- Control of insulin expression so it gives a signal for blood glucose regulation.
- X chromosome inactivation in female mammals to prevent an "overdose" of the genes it contains.
- Cyclin expression levels control progression through the eukaryotic cell cycle.

More generally, gene regulation gives the cell control over all structure and function, and is the basis for cellular differentiation, morphogenesis and the versatility and adaptability of any organism.

Any step of gene expression may be modulated, from the DNA-RNA transcription step to post-translational modification of a protein. The stability of the final gene product, whether it is RNA or protein, also contributes to the expression level of the gene—an unstable product results in a low expression level. In general gene expression is regulated through changes in the number and type of interactions between molecules that collectively influence transcription of DNA and translation of RNA.

Numerous terms are used to describe types of genes depending on how they are regulated; these include:

- A constitutive gene is a gene that is transcribed continually as opposed to a facultative gene, which is only transcribed when needed.
- A *housekeeping gene* is typically a constitutive gene that is transcribed at a relatively constant level. The housekeeping gene's products are typically needed for maintenance of the cell. It is generally assumed that their expression is unaffected by experimental conditions. Examples include actin, GAPDH and ubiquitin.
- A facultative gene is a gene only transcribed when needed as opposed to a constitutive gene.
- An inducible gene is a gene whose expression is either responsive to environmental change or dependent on the position in the cell cycle.

Transcriptional Regulation

Regulation of transcription can be broken down into three main routes of influence; genetic (direct interaction of a control factor with the gene), modulation interaction of a control factor with the transcription machinery and epigenetic (non-sequence changes in DNA structure that influence transcription).

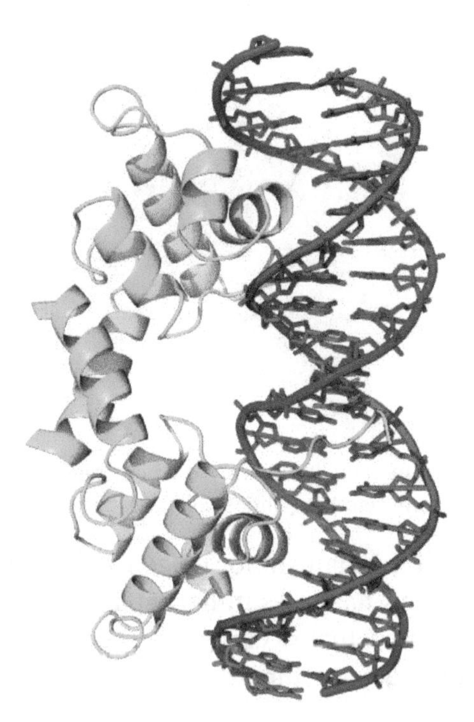

The lambda repressor transcription factor (green) binds as a dimer to major groove of DNA target (red and blue) and disables initiation of transcription. From PDB: 1LMB.

Direct interaction with DNA is the simplest and the most direct method by which a protein changes transcription levels. Genes often have several protein binding sites around the coding region with the specific function of regulating transcription. There are many classes of regulatory DNA binding sites known as enhancers, insulators and silencers. The mechanisms for regulating transcription are very varied, from blocking key binding sites on the DNA for RNA polymerase to acting as an activator and promoting transcription by assisting RNA polymerase binding.

The activity of transcription factors is further modulated by intracellular signals causing protein post-translational modification including phosphorylated, acetylated, or glycosylated. These changes influence a transcription factor's ability to bind, directly or indirectly, to promoter DNA, to recruit RNA polymerase, or to favor elongation of a newly synthesized RNA molecule.

The nuclear membrane in eukaryotes allows further regulation of transcription factors by the duration of their presence in the nucleus, which is regulated by reversible changes in their structure and by binding of other proteins. Environmental stimuli or endocrine signals may cause modification of regulatory proteins eliciting cascades of intracellular signals, which result in regulation of gene expression.

More recently it has become apparent that there is a significant influence of non-DNA-sequence specific effects on translation. These effects are referred to as epigenetic and involve the higher order structure of DNA, non-sequence specific DNA binding proteins and chemical modification of DNA. In general epigenetic effects alter the accessibility of DNA to proteins and so modulate transcription.

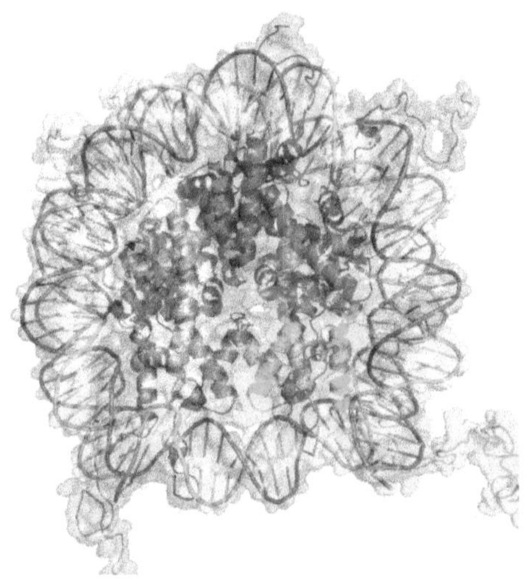

In eukaryotes, DNA is organized in form of nucleosomes. Note how the DNA (blue and green) is tightly wrapped around the protein core made of histone octamer (ribbon coils), restricting access to the DNA. From PDB: 1KX5.

DNA methylation is a widespread mechanism for epigenetic influence on gene expression and is seen in bacteria and eukaryotes and has roles in heritable transcription silencing and transcription regulation. In eukaryotes the structure of chromatin, controlled by the histone code, regulates access to DNA with significant impacts on the expression of genes in euchromatin and heterochromatin areas.

Post-transcriptional Regulation

In eukaryotes, where export of RNA is required before translation is possible, nuclear export is thought to provide additional control over gene expression. All transport in and out of the nucleus is via the nuclear pore and transport is controlled by a wide range of importin and exportin proteins.

Expression of a gene coding for a protein is only possible if the messenger RNA carrying the code survives long enough to be translated. In a typical cell, an RNA molecule is only stable if specifically protected from degradation. RNA degradation has particular importance in regulation of expression in eukaryotic cells where mRNA has to travel significant distances before being translated. In eukaryotes, RNA is stabilised by certain post-transcriptional modifications, particularly the 5' cap and poly-adenylated tail.

Intentional degradation of mRNA is used not just as a defence mechanism from foreign RNA (normally from viruses) but also as a route of mRNA *destabilisation*. If an mRNA molecule has a complementary sequence to a small interfering RNA then it is targeted for destruction via the RNA interference pathway.

Three Prime Untranslated Regions and MicroRNAs

Three prime untranslated regions (3'UTRs) of messenger RNAs (mRNAs) often contain regulatory sequences that post-transcriptionally influence gene expression. Such 3'-UTRs often contain both binding sites for microRNAs (miRNAs) as well as for regulatory proteins. By binding to specific sites within the 3'-UTR, miRNAs can decrease gene expression of various mRNAs by either inhibiting translation or directly causing degradation of the transcript. The 3'-UTR also may have silencer regions that bind repressor proteins that inhibit the expression of a mRNA.

The 3'-UTR often contains microRNA response elements (MREs). MREs are sequences to which miRNAs bind. These are prevalent motifs within 3'-UTRs. Among all regulatory motifs within the 3'-UTRs (e.g. including silencer regions), MREs make up about half of the motifs.

As of 2014, the miRBase web site, an archive of miRNA sequences and annotations, listed 28,645 entries in 233 biologic species. Of these, 1,881 miRNAs were in annotated human miRNA loci. miRNAs were predicted to have an average of about four hundred target mRNAs (affecting expression of several hundred genes). Freidman et al. estimate that >45,000 miRNA target sites within human mRNA 3'UTRs are conserved above background levels, and >60% of human protein-coding genes have been under selective pressure to maintain pairing to miRNAs.

Direct experiments show that a single miRNA can reduce the stability of hundreds of unique mRNAs. Other experiments show that a single miRNA may repress the production of hundreds of proteins, but that this repression often is relatively mild (less than 2-fold).

The effects of miRNA dysregulation of gene expression seem to be important in cancer. For instance, in gastrointestinal cancers, nine miRNAs have been identified as epigenetically altered and effective in down regulating DNA repair enzymes.

The effects of miRNA dysregulation of gene expression also seem to be important in neuropsychiatric disorders, such as schizophrenia, bipolar disorder, major depression, Parkinson's disease, Alzheimer's disease and autism spectrum disorders.

Translational Regulation

Direct regulation of translation is less prevalent than control of transcription or mRNA stability but is occasionally used. Inhibition of protein translation is a major target for toxins and antibiotics, so they can kill a cell by overriding its normal gene expression control. Protein synthesis inhibitors include the antibiotic neomycin and the toxin ricin.

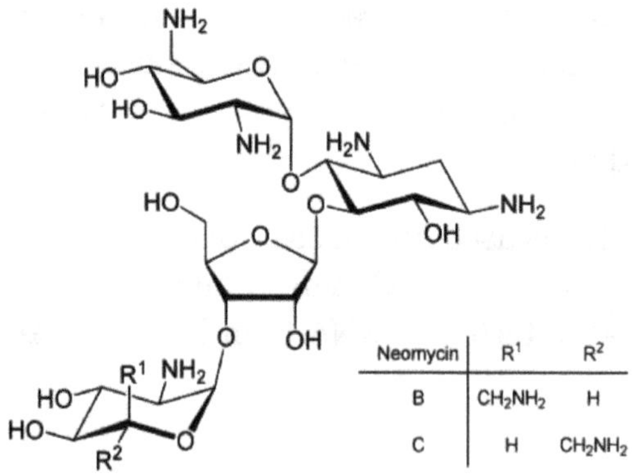

Neomycin	R[1]	R[2]
B	CH_2NH_2	H
C	H	CH_2NH_2

Neomycin is an example of a small molecule that reduces expression of all protein genes inevitably leading to cell death; it thus acts as an antibiotic.

Protein Degradation

Once protein synthesis is complete, the level of expression of that protein can be reduced by protein degradation. There are major protein degradation pathways in all prokaryotes and eukaryotes, of which the proteasome is a common component. An unneeded or damaged protein is often labeled for degradation by addition of ubiquitin.

Measurement

Measuring gene expression is an important part of many life sciences, as the ability to quantify the level at which a particular gene is expressed within a cell, tissue or organism can provide a lot of valuable information. For example, measuring gene expression can:

- Identify viral infection of a cell (viral protein expression).
- Determine an individual's susceptibility to cancer (oncogene expression).
- Find if a bacterium is resistant to penicillin (beta-lactamase expression).

Similarly, the analysis of the location of protein expression is a powerful tool, and this can be done on an organismal or cellular scale. Investigation of localization is particularly important for the study of development in multicellular organisms and as an indicator of protein function in single cells. Ideally, measurement of expression is done by detecting the final gene product (for many genes, this is the protein); however, it is often easier to detect one of the precursors, typically mRNA and to infer gene-expression levels from these measurements.

mRNA Quantification

Levels of mRNA can be quantitatively measured by northern blotting, which provides size and sequence information about the mRNA molecules. A sample of RNA is separated on an agarose gel and hybridized to a radioactively labeled RNA probe that is complementary to the target sequence. The radiolabeled RNA is then detected by an autoradiograph. Because the use of radioactive reagents makes the procedure time consuming and potentially dangerous, alternative labeling and detection methods, such as digoxigenin and biotin chemistries, have been developed. Perceived disadvantages of Northern blotting are that large quantities of RNA are required and that quantification may not be completely accurate, as it involves measuring band strength in an image of a gel. On the other hand, the additional mRNA size information from the Northern blot allows the discrimination of alternately spliced transcripts.

Another approach for measuring mRNA abundance is RT-qPCR. In this technique, reverse transcription is followed by quantitative PCR. Reverse transcription first generates a DNA template from the mRNA; this single-stranded template is called cDNA. The cDNA template is then amplified in the quantitative step, during which the fluorescence emitted by labeled hybridization probes or intercalating dyes changes as the DNA amplification process progresses. With a carefully constructed standard curve, qPCR can produce an absolute measurement of the number of copies of original mRNA, typically in units of copies per nanolitre of homogenized tissue or copies per cell. qPCR is very sensitive (detection of a single mRNA molecule is theoretically possible), but can be expensive depending on the type of reporter used; fluorescently labeled oligonucleotide probes are more expensive than non-specific intercalating fluorescent dyes.

For expression profiling, or high-throughput analysis of many genes within a sample, quantitative PCR may be performed for hundreds of genes simultaneously in the case of low-density arrays. A second approach is the hybridization microarray. A single array or "chip" may contain probes to determine transcript levels for every known gene in the genome of one or more organisms. Alternatively, "tag based" technologies like Serial analysis of gene expression (SAGE) and RNA-Seq, which can provide a relative measure of the cellular concentration of different mRNAs, can be used. An advantage of tag-based methods is the "open architecture", allowing for the exact measurement of any transcript, with a known or unknown sequence. Next-generation sequencing (NGS) such as RNA-Seq is another approach, producing vast quantities of sequence data that can be matched to a reference genome. Although NGS is comparatively time-consuming, expensive, and resource-intensive, it can identify single-nucleotide polymorphisms, splice-variants, and novel genes, and can also be used to profile expression in organisms for which little or no sequence information is available.

RNA Profiles

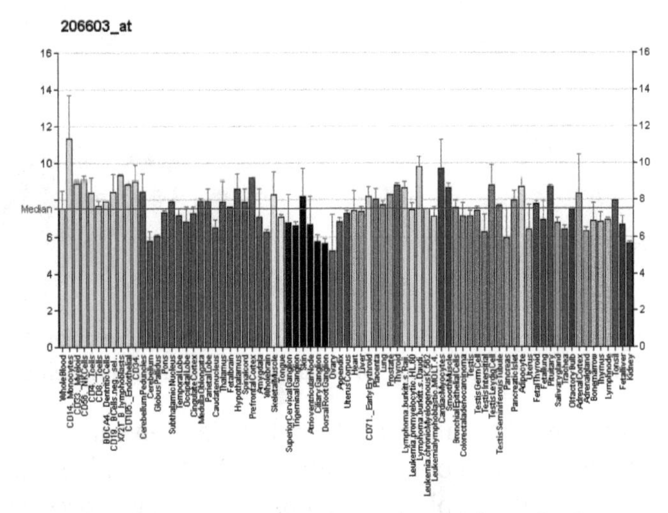

The RNA Expression profile of the GLUT4 Transporter (one of the main glucose transporters found in the human body.

Profiles like these are found for almost all proteins. They are generated by or-ganizations such as the Genomics Institute of the Novartis Research Foundation and the European Bioinformatics Institute. Additional information can be found by searching their databases. These profiles indicate the level of DNA expression (and hence RNA produced) of a certain protein in a certain tissue.

Protein Quantification

For genes encoding proteins, the expression level can be directly assessed by a number of methods with some clear analogies to the techniques for mRNA quantification.

The most commonly used method is to perform a Western blot against the protein of interest—this gives information on the size of the protein in addition to its identity. A sample (often cellular lysate) is separated on a polyacrylamide gel, transferred to a membrane and then probed with an antibody to the protein of interest. The antibody can either be conjugated to a fluorophore or to horseradish peroxidase for imaging and/or quantification. The gel-based nature of this assay makes quantification less accurate, but it has the advantage of being able to identify later modifications to the protein, for example proteolysis or ubiquitination, from changes in size.

Localisation

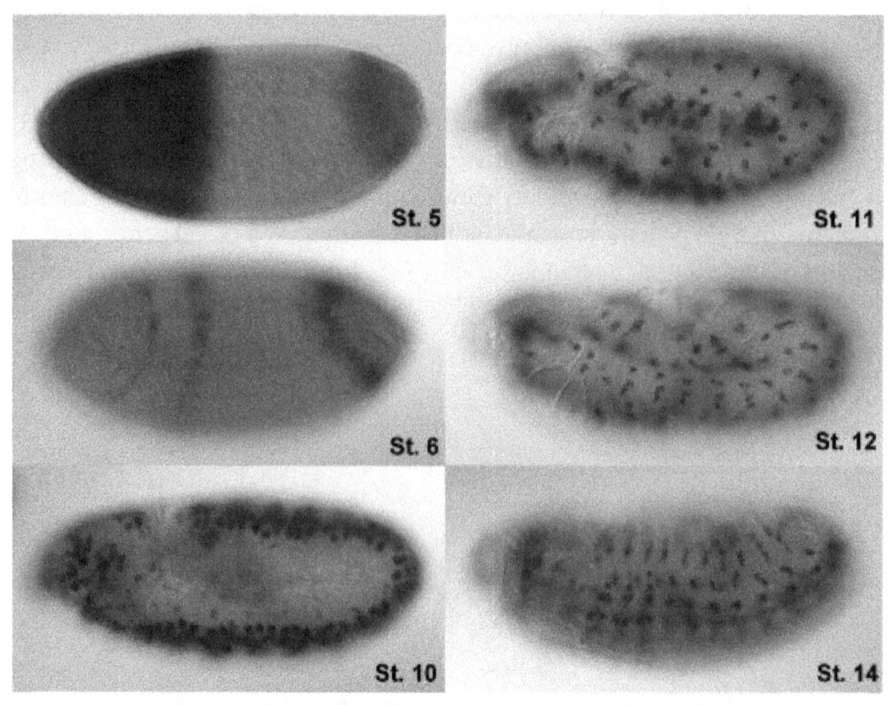

In situ-hybridization of Drosophila embryos at different developmental stages for the mRNA responsible for the expression of hunchback. High intensity of blue color marks places with high hunchback mRNA quantity.

Analysis of expression is not limited to quantification; localisation can also be determined. mRNA can be detected with a suitably labelled complementary mRNA strand and protein can be detected via labelled antibodies. The probed sample is then observed by microscopy to identify where the mRNA or protein is.

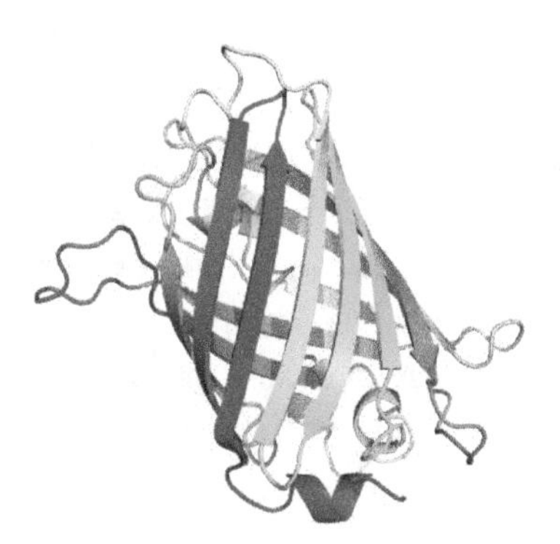

The three-dimensional structure of green fluorescent protein. The residues in the centre of the "barrel" are responsible for production of green light after exposing to higher energetic blue light. From PDB: 1EMA.

By replacing the gene with a new version fused to a green fluorescent protein (or similar) marker, expression may be directly quantified in live cells. This is done by imaging using a fluorescence microscope. It is very difficult to clone a GFP-fused protein into its native location in the genome without affecting expression levels so this method often cannot be used to measure endogenous gene expression. It is, however, widely used to measure the expression of a gene artificially introduced into the cell, for example via an expression vector. It is important to note that by fusing a target protein to a fluorescent reporter the protein's behavior, including its cellular localization and expression level, can be significantly changed.

The enzyme-linked immunosorbent assay works by using antibodies immobilised on a microtiter plate to capture proteins of interest from samples added to the well. Using a detection antibody conjugated to an enzyme or fluorophore the quantity of bound protein can be accurately measured by fluorometric or colourimetric detection. The detection process is very similar to that of a Western blot, but by avoiding the gel steps more accurate quantification can be achieved.

Expression System

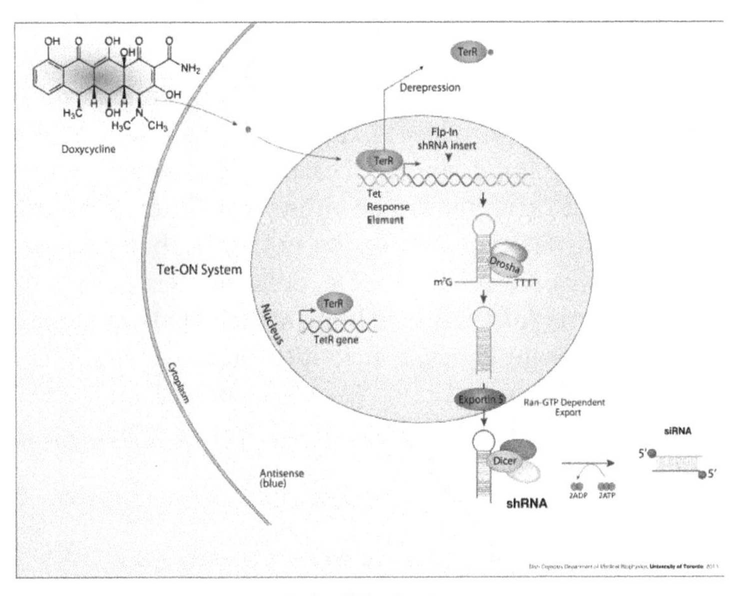

Tet-ON inducible shRNA system

An expression system is a system specifically designed for the production of a gene product of choice. This is normally a protein although may also be RNA, such as tRNA or a ribozyme. An expression system consists of a gene, normally encoded by DNA, and the molecular machinery required to transcribe the DNA into mRNA and translate the mRNA into protein using the reagents provided. In the broadest sense this includes every living cell but the term is more normally used to refer to expression as a laboratory tool. An expression system is therefore often artificial in some manner. Expression systems are, however, a fundamentally natural process. Viruses are an excellent example where they replicate by using the host cell as an expression system for the viral proteins and genome.

Inducible Expression

Doxycycline is also used in "Tet-on" and "Tet-off" tetracycline controlled transcriptional activation to regulate transgene expression in organisms and cell cultures.

In Nature

In addition to these biological tools, certain naturally observed configurations of DNA (genes, promoters, enhancers, repressors) and the associated machinery itself are referred to as an expression system. This term is normally used in the case where a gene or set of genes is switched on under well defined conditions, for example, the simple repressor switch expression system in Lambda phage and the lac operator system in bacteria. Several natural expression systems are directly used or modified and used for artificial expression systems such as the Tet-on and Tet-off expression system.

Gene Networks

Genes have sometimes been regarded as nodes in a network, with inputs being proteins such as transcription factors, and outputs being the level of gene expression. The node itself performs a function, and the operation of these functions have been interpreted as performing a kind of information processing within cells and determines cellular behavior.

Gene networks can also be constructed without formulating an explicit causal model. This is often the case when assembling networks from large expression data sets. Covariation and correlation of expression is computed across a large sample of cases and measurements (often transcriptome or proteome data). The source of variation can be either experimental or natural (observational). There are several ways to construct gene expression networks, but one common approach is to compute a matrix of all pair-wise correlations of expression across conditions, time points, or individuals and convert the matrix (after thresholding at some cut-off value) into a graphical representation in which nodes represent genes, transcripts, or proteins and edges connecting these nodes represent the strength of association. Weighted correlation network analysis involves weighted networks defined by soft-thresholding the pairwise correlations among variables (e.g. measures of transcript abundance).

Techniques and Tools

The following experimental techniques are used to measure gene expression and are listed in roughly chronological order, starting with the older, more established technologies. They are divided into two groups based on their degree of multiplexity.

- Low-to-mid-plex techniques:
    - Reporter gene
    - Northern blot
    - Western blot
    - Fluorescent in situ hybridization
    - Reverse transcription PCR
- Higher-plex techniques:
    - SAGE
    - DNA microarray
    - Tiling array
    - RNA-Seq

Translation (Biology)

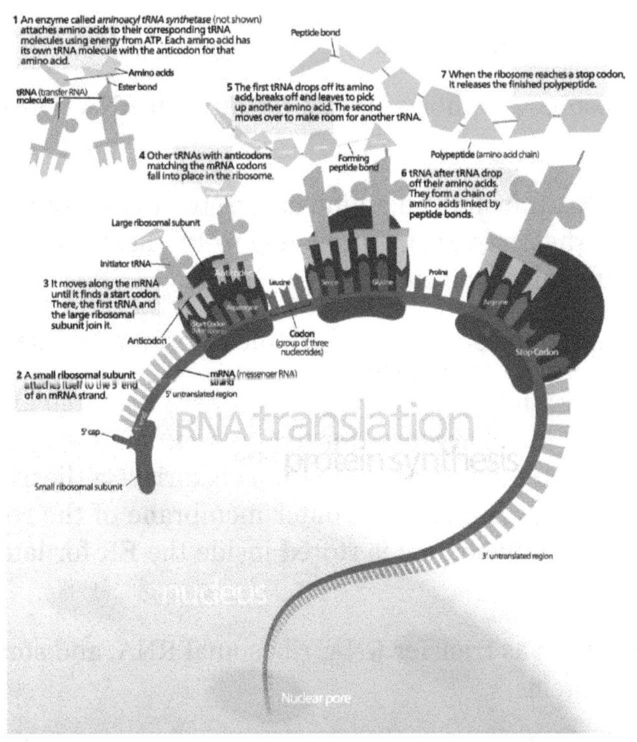

Overview of the translation of eukaryotic messenger RNA

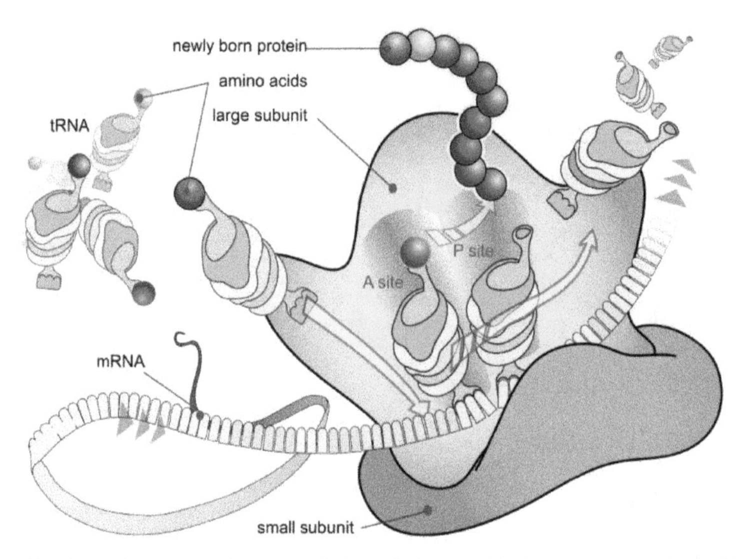

Diagram showing the translation of mRNA and the synthesis of proteins by a ribosome

In molecular biology and genetics, translation is the process in which cellular ribosomes create proteins.

In translation, messenger RNA (mRNA)—produced by transcription from DNA—is decoded by a ribosome to produce a specific amino acid chain, or polypeptide. The polypeptide later folds into an active protein and performs its functions in the cell. The ribosome facilitates decoding by inducing the binding of complementary tRNA anticodon sequences to mRNA codons. The tRNAs carry specific amino acids that are chained together into a polypeptide as the mRNA passes through and is "read" by the ribosome. The entire process is a part of gene expression.

In brief, translation proceeds in three phases:

1. Initiation: The ribosome assembles around the target mRNA. The first tRNA is attached at the start codon.

2. Elongation: The tRNA transfers an amino acid to the tRNA corresponding to the next codon. The ribosome then moves (*translocates*) to the next mRNA codon to continue the process, creating an amino acid chain.

3. Termination: When a stop codon is reached, the ribosome releases the polypeptide.

In bacteria, translation occurs in the cell's cytoplasm, where the large and small subunits of the ribosome bind to the mRNA. In eukaryotes, translation occurs in the cytosol or across the membrane of the endoplasmic reticulum in a process called vectorial synthesis. In many instances, the entire ribosome/mRNA complex binds to the outer membrane of the rough endoplasmic reticulum (ER); the newly created polypeptide is stored inside the ER for later vesicle transport and secretion outside of the cell.

Many of transcribed RNA, such as transfer RNA, ribosomal RNA, and small nuclear RNA, do not undergo translation into proteins.

A number of antibiotics act by inhibiting translation. These include anisomycin, cycloheximide, chloramphenicol, tetracycline, streptomycin, erythromycin, and puromycin. Prokaryotic ribosomes have a different structure from that of eukaryotic ribosomes, and thus antibiotics can specifically target bacterial infections without any harm to a eukaryotic host's cells.

Basic Mechanisms

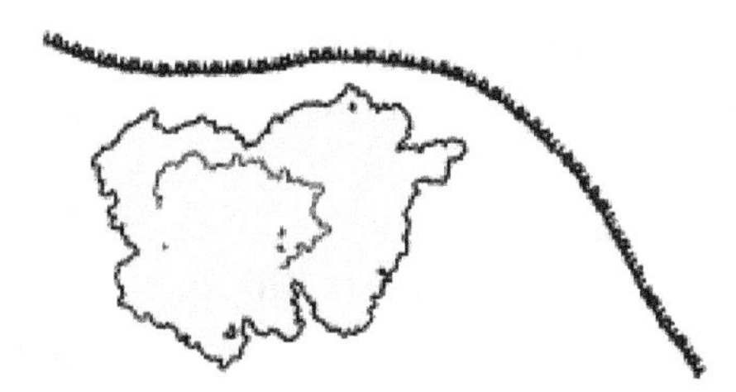

A ribosome translating a protein that is secreted into the endoplasmic reticulum. tRNAs are colored dark blue.

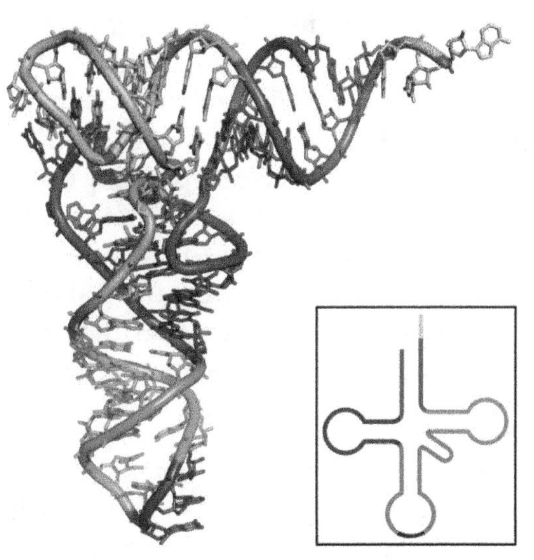

Tertiary structure of tRNA. *CCA tail* in orange, *Acceptor stem* in purple, *D arm* in red, *Anticodon arm* in blue with *Anticodon* in black, *T arm* in green.

The basic process of protein production is addition of one amino acid at a time to the end of a protein. This operation is performed by a ribosome. The choice of amino acid type to add is determined by an mRNA molecule. Each amino acid added is matched to a three nucleotide subsequence of the mRNA. For each such triplet possible, the corresponding amino acid is accepted. The successive amino acids added to the chain are matched to successive nucleotide triplets in the mRNA. In this way the sequence of nucleotides in the template mRNA chain determines the sequence of amino acids in the generated amino acid chain. Addition of an amino acid occurs at the C-terminus of the peptide and thus translation is said to be amino-to-carboxyl directed.

The mRNA carries genetic information encoded as a ribonucleotide sequence from the chromosomes to the ribosomes. The ribonucleotides are "read" by translational machinery in a sequence of nucleotide triplets called codons. Each of those triplets codes for a specific amino acid.

The ribosome molecules translate this code to a specific sequence of amino acids. The ribosome is a multisubunit structure containing rRNA and proteins. It is the "factory" where amino acids are assembled into proteins. tRNAs are small noncoding RNA chains (74-93 nucleotides) that transport amino acids to the ribosome. tRNAs have a site for amino acid attachment, and a site called an anticodon. The anticodon is an RNA triplet complementary to the mRNA triplet that codes for their cargo amino acid.

Aminoacyl tRNA synthetases (enzymes) catalyze the bonding between specific tRNAs and the amino acids that their anticodon sequences call for. The product of this reaction is an aminoacyl-tRNA. This aminoacyl-tRNA is carried to the ribosome by EF-Tu, where mRNA codons are matched through complementary base pairing to specific tRNA anticodons. Aminoacyl-tRNA synthetases that mispair tRNAs with the wrong amino acids can produce mischarged aminoacyl-tRNAs, which can result in inappropriate amino acids at the respective position in protein. This "mistranslation" of the genetic code naturally occurs at low levels in most organisms, but certain cellular environments cause an increase in permissive mRNA decoding, sometimes to the benefit of the cell.

The ribosome has three sites for tRNA to bind. They are the aminoacyl site (abbreviated A), the peptidyl site (abbreviated P) and the exit site (abbreviated E). With respect to the mRNA, the three sites are oriented 5' to 3' E-P-A, because ribosomes move toward the 3' end of mRNA. The A site binds the incoming tRNA with the complementary codon on the mRNA. The P site holds the tRNA with the growing polypeptide chain. The E site holds the tRNA without its amino acid. When an aminoacyl-tRNA initially binds to its corresponding codon on the mRNA, it is in the A site. Then, a peptide bond forms between the amino acid of the tRNA in the A site and the amino acid of the charged tRNA in the P site. The growing polypeptide chain is transferred to the tRNA in the A site. Translocation occurs, moving the tRNA in the P site, now without an amino acid, to the E site; the tRNA that was in the A site, now charged with the polypeptide chain, is moved to the P site. The tRNA in the E site leaves and another aminoacyl-tRNA enters the A site to repeat the process.

After the new amino acid is added to the chain, and after the mRNA is released out of the nucleus and into the ribosome's core, the energy provided by the hydrolysis of a GTP bound to the translocase EF-G (in prokaryotes) and eEF-2 (in eukaryotes) moves the ribosome down one codon towards the 3' end. The energy required for translation of proteins is significant. For a protein containing n amino acids, the number of high-energy phosphate bonds required to translate it is $4n$-1. The rate of translation varies; it is significantly higher in prokaryotic cells (up to 17-21 amino acid residues per second) than in eukaryotic cells (up to 6-9 amino acid residues per second).

In activation, the correct amino acid is covalently bonded to the correct transfer RNA (tRNA). The amino acid is joined by its carboxyl group to the 3' OH of the tRNA by an ester bond. When the tRNA has an amino acid linked to it, it is termed "charged". Initiation involves the small subunit of the ribosome binding to the 5' end of mRNA with the help of initiation factors (IF). Termination of the polypeptide happens when the A site of the ribosome faces a stop codon (UAA, UAG, or UGA). No tRNA can recognize or bind to this codon. Instead, the stop codon induces the binding of a release factor protein that prompts the disassembly of the entire ribosome/mRNA complex.

The process of translation is highly regulated in both eukaryotic and prokaryotic organisms. Regulation of translation can impact the global rate of protein synthesis which is closely coupled to the metabolic and proliferative state of a cell. In addition, recent work has revealed that genetic differences and their subsequent expression as mRNAs can also impact translation rate in an RNA-specific manner.

Genetic Code

Whereas other aspects such as the 3D structure, called tertiary structure, of protein can only be predicted using sophisticated algorithms, the amino acid sequence, called primary structure, can be determined solely from the nucleic acid sequence with the aid of a translation table.

This approach may not give the correct amino acid composition of the protein, in particular if unconventional amino acids such as selenocysteine are incorporated into the protein, which is coded for by a conventional stop codon in combination with a downstream hairpin (SElenoCysteine Insertion Sequence, or SECIS).

There are many computer programs capable of translating a DNA/RNA sequence into a protein sequence. Normally this is performed using the Standard Genetic Code, however, few programs can handle all the "special" cases, such as the use of the alternative initiation codons. For instance, the rare alternative start codon CTG codes for Methionine when used as a start codon, and for Leucine in all other positions.

Example: Condensed translation table for the Standard Genetic Code (from the NCBI Taxonomy webpage).

AAs = FFLLSSSSYY**CC*WLLLLPPPPHHQQRRRRIIIMTTTTNNKKSSRRVVVVAAAAD-DEEGGGG

Starts = ---M---------------M---------------M----------------------------

Base1 = TTTTTTTTTTTTTTTTCCCCCCCCCCCCCCCCAAAAAAAAAAAAAAAAGGGGGGGG-GGGGGGGG

Base2 = TTTTCCCCAAAAGGGGTTTTCCCCAAAAGGGGTTTTCCCCAAAAGGGGTTTTC-CCCAAAAGGGG

Base3 = TCAGTCAGTCAGTCAGTCAGTCAGTCAGTCAGTCAGTCAGTCAGTCAGTCAGT-CAGTCAGTCAG

Translation Tables

Even when working with ordinary eukaryotic sequences such as the Yeast genome, it is often desired to be able to use alternative translation tables—namely for translation of the mitochondrial genes. Currently the following translation tables are defined by the NCBI Taxonomy Group for the translation of the sequences in GenBank:

1: The Standard

2: The Vertebrate Mitochondrial Code

3: The Yeast Mitochondrial Code

4: The Mold, Protozoan, and Coelenterate Mitochondrial Code and the Mycoplasma/Spiroplasma Code

5: The Invertebrate Mitochondrial Code

6: The Ciliate, Dasycladacean and Hexamita Nuclear Code

9: The Echinoderm and Flatworm Mitochondrial Code

10: The Euplotid Nuclear Code

11: The Bacterial and Plant Plastid Code

12: The Alternative Yeast Nuclear Code

13: The Ascidian Mitochondrial Code

14: The Alternative Flatworm Mitochondrial Code

15: Blepharisma Nuclear Code

16: Chlorophycean Mitochondrial Code

21: Trematode Mitochondrial Code

22: Scenedesmus obliquus mitochondrial Code

23: Thraustochytrium Mitochondrial Code

Basic Mechanism of Translation (Biology)

Ribosome

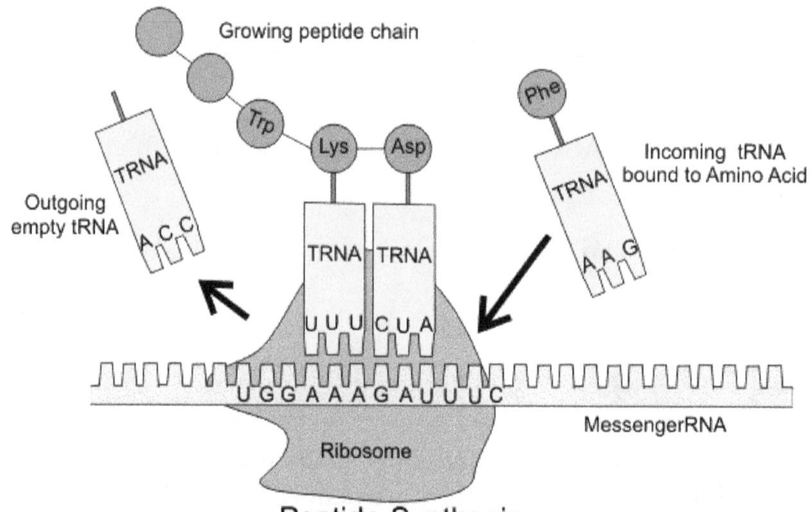

The ribosome assembles polymeric protein molecules whose sequence is controlled by the sequence of mes-senger RNA molecules. This is required by all living cells and associated viruses.

The ribosome is a complex molecular machine, found within all living cells, that serves as the site of biological protein synthesis (translation). Ribosomes link amino acids together in the order specified by messenger RNA (mRNA) molecules. Ribosomes consist of two major components: the small ribosomal subunit, which reads the RNA, and the large subunit, which joins amino acids to form a polypeptide chain. Each subunit is composed of one or more ribosomal RNA (rRNA) molecule and a variety of proteins. The ribosomes and associated molecules are also known as the *translational apparatus*.

The sequence of DNA, which encodes the sequence of the amino acids in a protein, is copied into a messenger RNA chain. It may be copied many times into RNA chains. Ribosomes can bind to a messenger RNA chain and use its sequence for determining the correct sequence of amino acids. Amino acids are selected, collected, and carried to the ribosome by transfer RNA (tRNA) molecules, which enter one part of the ribosome and bind to the messenger RNA chain. It is during this binding that the correct translation of nucleic acid sequence to amino acid sequence occurs. For each coding triplet in the messenger RNA there is a distinct transfer RNA that matches and which carries the correct amino acid for that coding triplet. The attached amino acids are then linked together by another part of the ribosome. Once the protein is produced, it can then fold to produce a specific functional three-dimensional structure although during synthesis some proteins start folding into their correct form.

A ribosome is made from complexes of RNAs and proteins and is therefore a ribonucleoprotein. Each ribosome is divided into two subunits: 1. a smaller subunit which binds to a larger subunit and the mRNA pattern, and 2. a larger subunit which binds to the tRNA, the amino acids, and the smaller subunit. When a ribosome finishes reading an mRNA molecule, these two subunits split apart. Ribosomes are ribozymes, because the catalytic peptidyl transferase activity that links amino acids together is performed by the ribosomal RNA. Ribosomes are often embedded in the intracellular membranes that make up the rough endoplasmic reticulum.

Ribosomes from bacteria, archaea and eukaryotes (the three domains of life on Earth) resemble each other to a remarkable degree, evidence of a common origin. They differ in their size, sequence, structure, and the ratio of protein to RNA. The differences in structure allow some antibiotics to kill bacteria by inhibiting their ribosomes, while leaving human ribosomes unaffected. In bacteria and archaea, more than one ribosome may move along a single mRNA chain at one time, each "reading" its sequence and producing a corresponding protein molecule.

The ribosomes in the mitochondria of eukaryotic cells (mitoribosomes), are produced from mitochondrial genes, and functionally resemble many features of those in bacteria, reflecting the likely evolutionary origin of mitochondria.

Discovery

Ribosomes were first observed in the mid-1950s by Romanian cell biologist George Emil Palade using an electron microscope as dense particles or granules for which, in 1974, he would win a Nobel Prize. The term "ribosome" was proposed by scientist Richard B. Roberts in 1958:

During the course of the symposium a semantic difficulty became apparent. To some of the participants, "microsomes" mean the ribonucleoprotein particles of the microsome fraction contaminated by other protein and lipid material; to others, the microsomes consist of protein and lipid contaminated by particles. The phrase "microsomal particles" does not seem adequate, and "ribonucleoprotein particles of the microsome fraction" is much too awkward. During the meeting, the word "ribosome" was suggested, which has a very satisfactory name and a pleasant sound. The present confusion would be eliminated if "ribosome" were adopted to designate ribonucleoprotein particles in sizes ranging from 35 to 100S.

—Albert, Microsomal Particles and Protein Synthesis

Albert Claude, Christian de Duve, and George Emil Palade were jointly awarded the Nobel Prize in Physiology or Medicine, in 1974, for the discovery of the ribosomes. The Nobel Prize in Chemistry 2009 was awarded to Venkatraman Ramakrishnan, Thomas A. Steitz and Ada E. Yonath for determining the detailed structure and mechanism of the ribosome.

Structure

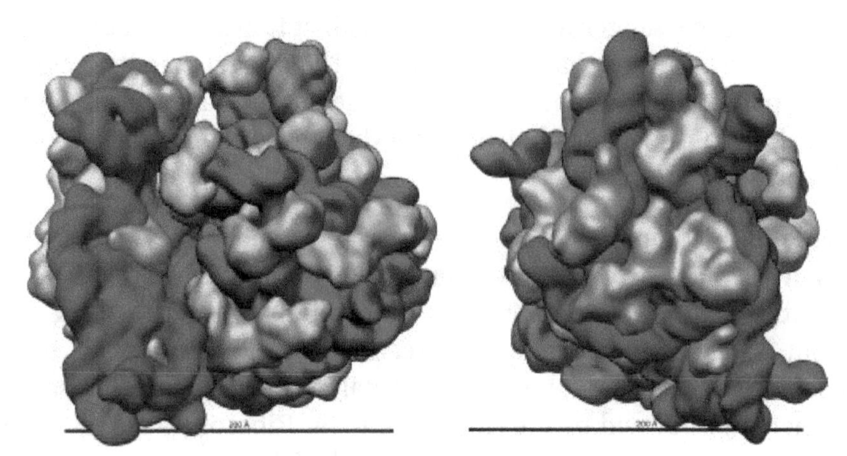

Large (red) and small (blue) subunit fit together.

The ribosome is responsible for the synthesis of proteins in cells and is found in all cellular organisms. It serves to convert the instructions found in messenger RNA (mRNA, which itself is made from instructions in DNA) into the chains of amino-acids that make up proteins.

The ribosome is a cellular machine which is highly complex. It is made up of dozens of distinct proteins (the exact number varies slightly between species) as well as a few specialized RNA molecules known as ribosomal RNA (rRNA). Note – these rRNAs do not carry instructions to make specific proteins like mRNAs. The ribosomal proteins and rRNAs are arranged into two distinct ribosomal pieces of different size, known generally as the large and small subunit of the ribosome. Ribosomes consist of two subunits that fit together (Figure 2) and work as one to translate the mRNA into a polypeptide chain during protein synthesis (Figure 1). Because they are formed from two subunits of non-equal size, they are slightly longer in the axis than in diameter. Prokaryotic ribosomes are around 20 nm (200 Å) in diameter and are composed of 65% rRNA and 35% ribosomal proteins. Eukaryotic ribosomes are between 25 and 30 nm (250–300 Å) in diameter with an rRNA-to-protein ratio that is close to 1. Bacterial ribosomes are composed of one or two rRNA strands. Eukaryotic ribosomes contain one or three very large rRNA molecules and multiple smaller protein molecules. Crystallographic work has shown that there are no ribosomal proteins close to the reaction site for polypeptide synthesis. This proves that the protein components of ribosomes do not directly participate in peptide bond formation catalysis, but rather suggests that these proteins act as a scaffold that may enhance the ability of rRNA to synthesize protein.

The ribosomal subunits of prokaryotes and eukaryotes are quite similar.

The unit of measurement is the Svedberg unit, a measure of the rate of sedimentation in centrifugation rather than size. This accounts for why fragment names do not add up: for example, prokaryotic 70S ribosomes are made of 50S and 30S subunits.

Prokaryotes have 70S ribosomes, each consisting of a small (30S) and a large (50S) subunit. Their small subunit has a 16S RNA subunit (consisting of 1540 nucleotides) bound to 21 proteins. The large subunit is composed of a 5S RNA subunit (120 nucleotides), a 23S RNA subunit (2900 nucleotides) and 31 proteins. Affinity label for the tRNA binding sites on the E. coli ribosome allowed the identification of A and P site proteins most likely associated with the peptidyltransferase activity; labelled proteins are L27, L14, L15, L16, L2; at least L27 is located at the donor site, as shown by E. Collatz and A.P. Czernilofsky. Additional research has demonstrated that the S1 and S21 proteins, in association with the 3'-end of 16S ribosomal RNA, are involved in the initiation of translation.

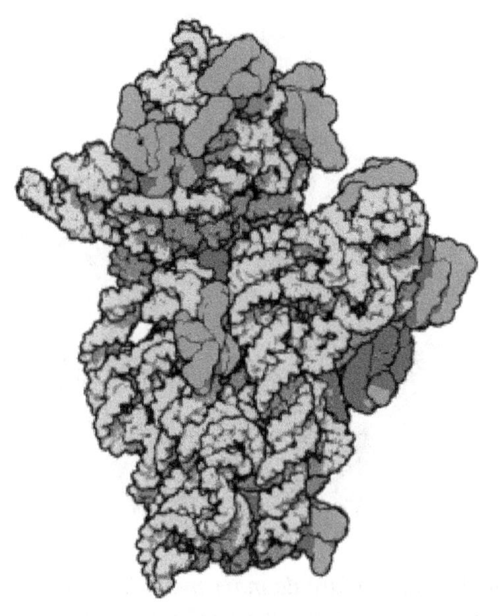

Atomic structure of the 30S Subunit from *Thermus thermophilus*. Proteins are shown in blue and the single RNA chain in orange.

Eukaryotes have 80S ribosomes, each consisting of a small (40S) and large (60S) subunit. Their 40S subunit has an 18S RNA (1900 nucleotides) and 33 proteins. The large subunit is composed of a 5S RNA (120 nucleotides), 28S RNA (4700 nucleotides), a 5.8S RNA (160 nucleotides) subunits and 46 proteins. During 1977, Czernilofsky published research that used affinity labeling to identify tRNA-binding sites on rat liver ribosomes. Several proteins, including L32/33, L36, L21, L23, L28/29 and L13 were implicated as being at or near the peptidyl transferase center.

The ribosomes found in chloroplasts and mitochondria of eukaryotes also consist of large and small subunits bound together with proteins into one 70S particle. These organelles are believed to be descendants of bacteria and, as such, their ribosomes are similar to those of bacteria.

The various ribosomes share a core structure, which is quite similar despite the large differences in size. Much of the RNA is highly organized into various tertiary structural motifs, for example pseudoknots that exhibit coaxial stacking. The extra RNA in the larger ribosomes is in several long continuous insertions, such that they form loops out of the core structure without disrupting or changing it. All of the catalytic activity of the ribosome is carried out by the RNA; the proteins reside on the surface and seem to stabilize the structure.

The differences between the bacterial and eukaryotic ribosomes are exploited by pharmaceutical chemists to create antibiotics that can destroy a bacterial infection without harming the cells of the infected person. Due to the differences in their structures, the bacterial 70S ribosomes are vulnerable to these antibiotics while the eukaryotic 80S ribosomes are not. Even though mitochondria possess ribosomes similar to the bacterial ones, mitochondria are not affected by these antibiotics because they are surrounded by a double membrane that does not easily admit these antibiotics into the organelle.

High-resolution Structure

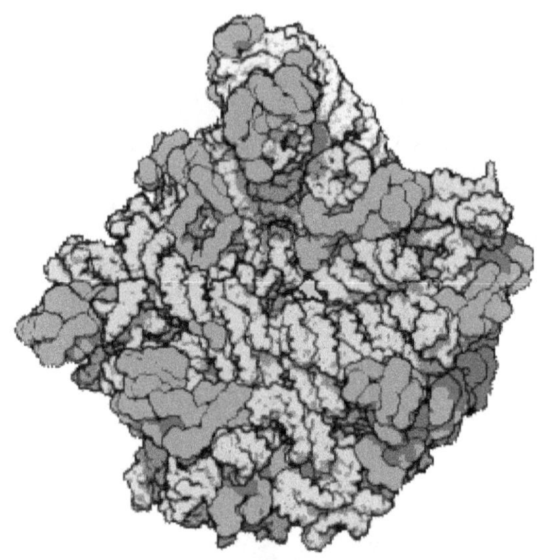

Atomic structure of the 50S Subunit from *Haloarcula marismortui*. Proteins are shown in blue and the two RNA chains in orange and yellow. The small patch of green in the center of the subunit is the active site.

The general molecular structure of the ribosome has been known since the early 1970s. In the early 2000s, the structure has been achieved at high resolutions, of the order of a few Å.

The first papers giving the structure of the ribosome at atomic resolution were published almost simultaneously in late 2000. The 50S (large prokaryotic) subunit was determined from the archaeon *Haloarcula marismortui* and the bacterium *Deinococcus radiodurans*, and the structure of the 30S subunit was determined from *Thermus thermophilus*. These structural studies were awarded the Nobel Prize in Chemistry in 2009. In May 2001 these coordinates were used to reconstruct the entire *T. thermophilus* 70S particle at 5.5 Å resolution.

Two papers were published in November 2005 with structures of the *Escherichia coli* 70S ribosome. The structures of a vacant ribosome were determined at 3.5-Å resolution using x-ray crystallography. Then, two weeks later, a structure based on cryo-electron microscopy was published, which depicts the ribosome at 11–15Å resolution in the act of passing a newly synthesized protein strand into the protein-conducting channel.

The first atomic structures of the ribosome complexed with tRNA and mRNA molecules were solved by using X-ray crystallography by two groups independently, at 2.8 Å and at 3.7 Å. These structures allow one to see the details of interactions of the *Thermus thermophilus* ribosome with mRNA and with tRNAs bound at classical ribosomal sites. Interactions of the ribosome with long mRNAs containing Shine-Dalgarno sequences were visualized soon after that at 4.5- to 5.5-Å resolution.

In 2011, the first complete atomic structure of the eukaryotic 80S ribosome from the yeast *Saccharomyces cerevisiae* was obtained by crystallography. The model reveals the architecture of eukaryote-specific elements and their interaction with the universally conserved core. At the same time, the complete model of a eukaryotic 40S ribosomal structure in *Tetrahymena thermophila* was published and described the structure of the 40S subunit, as well as much about the 40S subunit's interaction with eIF1 during translation initiation. Similarly, the eukaryotic 60S subunit structure was also determined from *Tetrahymena thermophila* in complex with eIF6.

Function

Ribosomes are a cell structure that makes protein. Protein is needed for many cell functions such as repairing damage or directing chemical processes. Ribosomes can be found floating within the cytoplasm or attached to the endoplasmic reticulum.

Translation

Ribosomes are the workplaces of protein biosynthesis, the process of translating mRNA into protein. The mRNA comprises a series of codons that dictate to the ribosome the sequence of the amino acids needed to make the protein. Using the mRNA as a template, the ribosome traverses each codon (3 nucleotides) of the mRNA, pairing it with the appropriate amino acid provided by an aminoacyl-tRNA. Aminoacyl-tRNA contains a complementary anticodon on one end and the appropriate amino acid on the other. For fast and accurate recognition of the appropriate tRNA, the ribosome utilizes large conformational changes (conformational proofreading) . The small ribosomal subunit, typically bound to an aminoacyl-tRNA containing the amino acid methionine, binds to an AUG codon on the mRNA and recruits the large ribosomal subunit. The ribosome contains three RNA binding sites, designated A, P and E. The A site binds an aminoacyl-tRNA; the P site binds a peptidyl-tRNA (a tRNA bound to the peptide being synthesized); and the E site binds a free tRNA before it exits the ribosome. Protein synthesis begins at a start codon AUG near the 5' end of the mRNA. mRNA binds to the P site of the ribosome first. The ribosome is able to identify the start codon by use of the Shine-Dalgarno sequence of the mRNA in prokaryotes and Kozak box in eukaryotes.

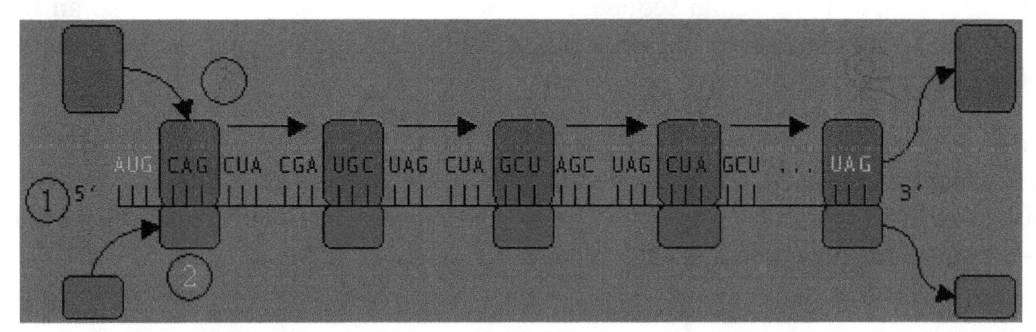

Translation of mRNA (1) by a ribosome (2)(shown as small and large subunits) into a polypeptide chain (3). The ribosome begins at the start codon of RNA (AUG) and ends at the stop codon (UAG).

Although catalysis of the peptide bond involves the C2 hydroxyl of RNA's P-site adenosine in a proton shuttle mechanism, other steps in protein synthesis (such as translocation) are caused by changes in protein conformations. Since their catalytic core is made of RNA, ribosomes are classified as "ribozymes," and it is thought that they might be remnants of the RNA world.

In Figure 5, both ribosomal subunits (small and large) assemble at the start codon (towards the 5' end of the RNA). The ribosome uses RNA that matches the current codon (triplet) on the mRNA to append an amino acid to the polypeptide chain. This is done for each triplet on the RNA, while the ribosome moves towards the 3' end of the mRNA. Usually in bacterial cells, several ribosomes are working parallel on a single RNA, forming what is called a *polyribosome* or *polysome*.

Addition of Translation-independent Amino Acids

Presence of a ribosome quality control protein Rqc2 is associated with mRNA-independent protein elongation. This elongation is a result of ribosomal addition (via tRNAs brought by Rqc2) of *CAT tails*: ribosomes extend the *C*-terminus of a stalled protein with random, translation-independent sequences of *a*lanines and *t*hreonines.

Ribosome Locations

Ribosomes are classified as being either "free" or "membrane-bound".

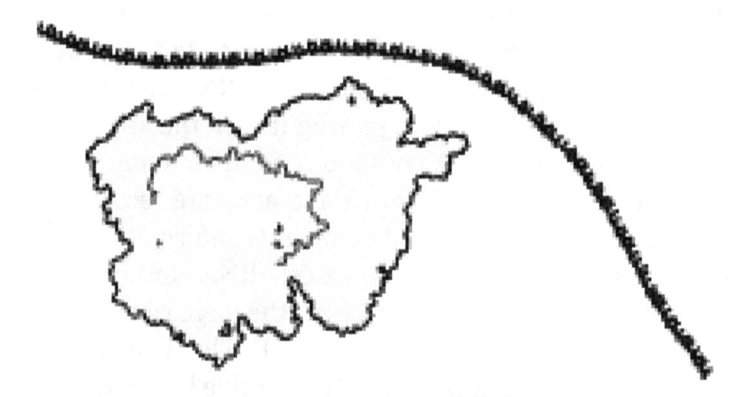

A ribosome translating a protein that is secreted into the endoplasmic reticulum.

Free and membrane-bound ribosomes differ only in their spatial distribution; they are identical in structure. Whether the ribosome exists in a free or membrane-bound state depends on the presence of an ER-targeting signal sequence on the protein being synthesized, so an individual ribosome might be membrane-bound when it is making one protein, but free in the cytosol when it makes another protein.

Ribosomes are sometimes referred to as organelles, but the use of the term *organelle* is often restricted to describing sub-cellular components that include a phospholipid membrane, which ribosomes, being entirely particulate, do not. For this reason, ribosomes may sometimes be described as "non-membranous organelles".

Free Ribosomes

Free ribosomes can move about anywhere in the cytosol, but are excluded from the cell nucleus and other organelles. Proteins that are formed from free ribosomes are released into the cytosol and used within the cell. Since the cytosol contains high concentrations of glutathione and is, therefore, a reducing environment, proteins containing disulfide bonds, which are formed from oxidized cysteine residues, cannot be produced within it.

Membrane-bound Ribosomes

When a ribosome begins to synthesize proteins that are needed in some organelles, the ribosome making this protein can become "membrane-bound". In eukaryotic cells this happens in a region of the endoplasmic reticulum (ER) called the "rough ER". The newly produced polypeptide chains are inserted directly into the ER by the ribosome undertaking vectorial synthesis and are then transported to their destinations, through the secretory pathway. Bound ribosomes usually produce proteins that are used within the plasma membrane or are expelled from the cell via *exocytosis*.

Biogenesis

In bacterial cells, ribosomes are synthesized in the cytoplasm through the transcription of multiple ribosome gene operons. In eukaryotes, the process takes place both in the cell cytoplasm and in the nucleolus, which is a region within the cell nucleus. The assembly process involves the coordinated function of over 200 proteins in the synthesis and processing of the four rRNAs, as well as assembly of those rRNAs with the ribosomal proteins.

Origin

The ribosome may have first originated in an RNA world, appearing as a self-replicating complex that only later evolved the ability to synthesize proteins when amino acids began to appear. Studies suggest that ancient ribosomes constructed solely of rRNA could have developed the ability to synthesize peptide bonds. In addition, evidence strongly points to ancient ribosomes as self-replicating complexes, where the rRNA in the ribosomes had informational, structural, and catalytic purposes because it could have coded for tRNAs and proteins needed for ribosomal self-replication. As amino acids gradually appeared in the RNA world under prebiotic conditions, their interactions with catalytic RNA would increase both the range and efficiency of function of catalytic RNA molecules. Thus, the driving force for the evolution of the ribosome from an ancient self-replicating machine into its current form as a translational machine may have been the selective pressure to incorporate proteins into the ribosome's self-replicating mechanisms, so as to increase its capacity for self-replication.

Heterogeneity

Heterogeneity in ribosome composition has been proposed to be involved in translational control of protein synthesis. Vincent Mauro and Gerald Edelman proposed ribosome filter hypothesis to explain the regulatory functions of ribosomes.

RNA Interference

Lentiviral delivery of designed shRNA's and the mechanism of RNA interference in mammalian cells.

RNA interference (RNAi) is a biological process in which RNA molecules inhibit gene expression, typically by causing the destruction of specific mRNA molecules. Historically, it was known by other names, including *co-suppression*, *post-transcriptional gene silencing* (PTGS), and *quelling*. Only after these apparently unrelated processes were fully understood did it become clear that they all described the RNAi phenomenon. Andrew Fire and Craig C. Mello shared the 2006 Nobel Prize in Physiology or Medicine for their work on RNA interference in the nematode worm *Caenorhabditis elegans*, which they published in 1998. Since the discovery of RNAi and its regulatory potentials, it has become evident that RNAi has immense potential in suppression of desired genes. RNAi is now known as precise, efficient, stable and better than antisense technology for gene suppression.

Two types of small ribonucleic acid (RNA) molecules – microRNA (miRNA) and small interfering RNA (siRNA) – are central to RNA interference. RNAs are the direct products of genes, and these small RNAs can bind to other specific messenger RNA (mRNA) molecules and either increase or decrease their activity, for example by preventing an mRNA from producing a protein. RNA interference has an important role in defending cells against parasitic nucleotide sequences – viruses and transposons. It also influences development.

The RNAi pathway is found in many eukaryotes, including animals, and is initiated by the enzyme Dicer, which cleaves long double-stranded RNA (dsRNA) molecules into short double-stranded fragments of ~20 nucleotide siRNAs. Each siRNA is unwound into two single-stranded RNAs (ssRNAs), the passenger strand and the guide strand. The passenger strand is degraded and the guide strand is incorporated into the RNA-induced silencing complex (RISC). The most well-studied outcome is post-transcriptional gene silencing, which occurs when the guide strand pairs with a complementary sequence in a messenger RNA molecule and induces cleavage by Argonaute, the catalytic component of the RISC complex. In some organisms, this process spreads systemically, despite the initially limited molar concentrations of siRNA.

RNAi is a valuable research tool, both in cell culture and in living organisms, because synthetic dsRNA introduced into cells can selectively and robustly induce suppression of specific genes of interest. RNAi may be used for large-scale screens that systematically shut down each gene in the cell, which can help to identify the components necessary for a particular cellular process or an event such as cell division. The pathway is also used as a practical tool in biotechnology, medicine and insecticides.

Cellular Mechanism

RNAi is an RNA-dependent gene silencing process that is controlled by the RNA-induced silencing complex (RISC) and is initiated by short double-stranded RNA molecules in a cell's cytoplasm, where they interact with the catalytic RISC component argonaute. When the dsRNA is exogenous (coming from infection by a virus with an RNA genome or laboratory manipulations), the RNA is imported directly into the cytoplasm and cleaved to short fragments by Dicer. The initiating dsRNA can also be endogenous (originating in the cell), as in pre-microRNAs expressed from RNA-coding genes in the genome. The primary transcripts from such genes are first processed to form the characteristic stem-loop structure of pre-miRNA in the nucleus, then exported to the cytoplasm. Thus, the two dsRNA pathways, exogenous and endogenous, converge at the RISC.

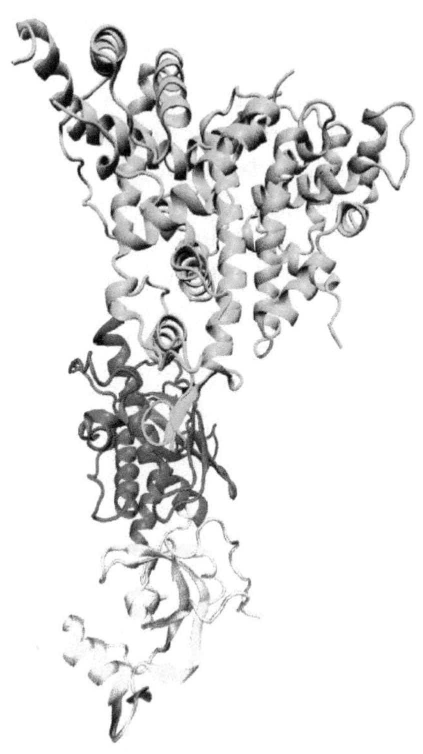

The dicer protein from *Giardia intestinalis,* which catalyzes the cleavage of dsRNA to siRNAs. The RNase domains are colored green, the PAZ domain yellow, the platform domain red, and the connector helix blue.

dsRNA Cleavage

Endogenous dsRNA initiates RNAi by activating the ribonuclease protein Dicer, which binds and cleaves double-stranded RNAs (dsRNAs) to produce double-stranded fragments of 20–25 base pairs with a 2-nucleotide overhang at the 3' end. Bioinformatics studies on the genomes of multiple organisms suggest this length maximizes target-gene specificity and minimizes non-specific effects. These short double-stranded fragments are called small interfering RNAs (siRNAs). These siRNAs are then separated into single strands and integrated into an active RISC complex. After integration into the RISC, siRNAs base-pair to their target mRNA and cleave it, thereby preventing it from being used as a translation template.

Exogenous dsRNA is detected and bound by an effector protein, known as RDE-4 in *C. elegans* and R2D2 in *Drosophila*, that stimulates dicer activity. This protein only binds long dsRNAs, but the mechanism producing this length specificity is unknown. This RNA-binding protein then facilitates the transfer of cleaved siRNAs to the RISC complex.

In *C. elegans* this initiation response is amplified through the synthesis of a population of 'secondary' siRNAs during which the dicer-produced initiating or 'primary' siRNAs are used as templates. These 'secondary' siRNAs are structurally distinct from dicer-produced siRNAs and appear to be produced by an RNA-dependent RNA polymerase (RdRP).

MicroRNA

MicroRNAs (miRNAs) are genomically encoded non-coding RNAs that help regulate gene expression, particularly during development. The phenomenon of RNA interference, broadly defined, includes the endogenously induced gene silencing effects of miRNAs as well as silencing triggered by foreign dsRNA. Mature miRNAs are structurally similar to siRNAs produced from exogenous dsRNA, but before reaching maturity, miRNAs must first undergo extensive post-transcriptional modification. A miRNA is expressed from a much longer RNA-coding gene as a primary transcript known as a *pri-miRNA* which is processed, in the cell nucleus, to a 70-nucleotide stem-loop structure called a *pre-miRNA* by the microprocessor complex. This complex consists of an RNase III enzyme called Drosha and a dsRNA-binding protein DGCR8. The dsRNA portion of this pre-miRNA is bound and cleaved by Dicer to produce the mature miRNA molecule that can be integrated into the RISC complex; thus, miRNA and siRNA share the same downstream cellular machinery. First, viral encoded miRNA was described in EBV. Thereafter, an increasing number of microRNAs have been described in viruses. VIRmiRNA is a comprehensive catalogue covering viral microRNA, their targets and anti-viral miRNAs.

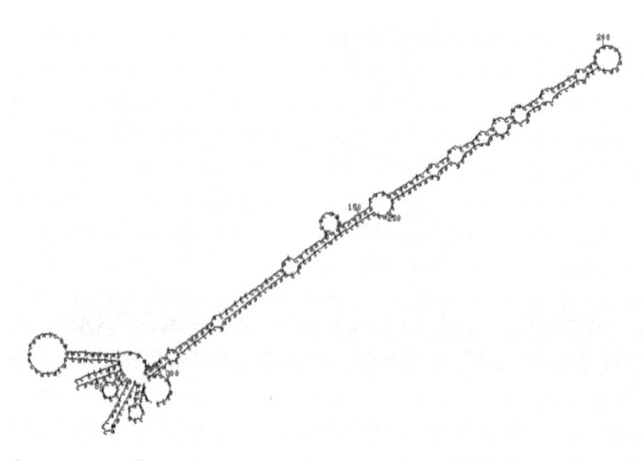

The stem-loop secondary structure of a pre-microRNA from *Brassica oleracea*.

siRNAs derived from long dsRNA precursors differ from miRNAs in that miRNAs, especially those in animals, typically have incomplete base pairing to a target and inhibit the translation of many different mRNAs with similar sequences. In contrast, siRNAs typically base-pair perfectly and induce mRNA cleavage only in a single, specific target. In *Drosophila* and *C. elegans*, miRNA and siRNA are processed by distinct argonaute proteins and dicer enzymes.

Three Prime Untranslated Regions and microRNAs

Three prime untranslated regions (3'UTRs) of messenger RNAs (mRNAs) often contain regulatory sequences that post-transcriptionally cause RNA interference. Such 3'-UTRs often contain both binding sites for microRNAs (miRNAs) as well as for regulatory proteins. By binding to specific sites within the 3'-UTR, miRNAs can decrease gene expression of various mRNAs by either inhibiting translation or directly causing degradation of the transcript. The 3'-UTR also may have silencer regions that bind repressor proteins that inhibit the expression of a mRNA.

The 3'-UTR often contains microRNA response elements (MREs). MREs are sequences to which miRNAs bind. These are prevalent motifs within 3'-UTRs. Among all regulatory motifs within the 3'-UTRs (e.g. including silencer regions), MREs make up about half of the motifs.

As of 2014, the miRBase web site, an archive of miRNA sequences and annotations, listed 28,645 entries in 233 biologic species. Of these, 1,881 miRNAs were in annotated human miRNA loci. miRNAs were predicted to have an average of about four hundred target mRNAs (affecting expression of several hundred genes). Friedman et al. estimate that >45,000 miRNA target sites within human mRNA 3'UTRs are conserved above background levels, and >60% of human protein-coding genes have been under selective pressure to maintain pairing to miRNAs.

Direct experiments show that a single miRNA can reduce the stability of hundreds of unique mRNAs. Other experiments show that a single miRNA may repress the production of hundreds of proteins, but that this repression often is relatively mild (less than 2-fold).

The effects of miRNA dysregulation of gene expression seem to be important in cancer. For instance, in gastrointestinal cancers, nine miRNAs have been identified as epigenetically altered and effective in down regulating DNA repair enzymes.

The effects of miRNA dysregulation of gene expression also seem to be important in neuropsychiatric disorders, such as schizophrenia, bipolar disorder, major depression, Parkinson's disease, Alzheimer's disease and autism spectrum disorders.

RISC Activation and Catalysis

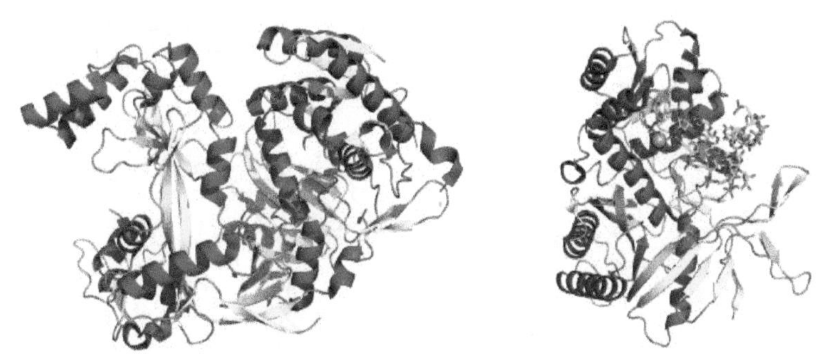

Left: A full-length argonaute protein from the archaea species *Pyrococcus furiosus*. *Right:* The PIWI domain of an argonaute protein in complex with double-stranded RNA.

The active components of an RNA-induced silencing complex (RISC) are endonucleases called argonaute proteins, which cleave the target mRNA strand complementary to their bound siRNA. As the fragments produced by dicer are double-stranded, they could each in theory produce a functional siRNA. However, only one of the two strands, which is known as the *guide strand*, binds the argonaute protein and directs gene silencing. The other *anti-guide strand* or *passenger strand* is degraded during RISC activation. Although it was first believed that an ATP-dependent helicase separated these two strands, the process proved to be ATP-independent and performed directly by the protein components of RISC. However, an *in vitro* kinetic analysis of RNAi in the presence and absence of ATP showed that ATP may be required to unwind and remove the cleaved mRNA strand from the RISC complex after catalysis. The guide strand tends to be the one whose 5' end is less stably paired to its complement, but strand selection is unaffected by the direction in which dicer cleaves the dsRNA before RISC incorporation. Instead, the R2D2 protein may serve as the differentiating factor by binding the more-stable 5' end of the passenger strand.

The structural basis for binding of RNA to the argonaute protein was examined by X-ray crystallography of the binding domain of an RNA-bound argonaute protein. Here, the phosphorylated 5' end of the RNA strand enters a conserved basic surface pocket and makes contacts through a divalent cation (an atom with two positive charges) such as magnesium and by aromatic stacking (a process that allows more than one atom to share an electron by passing it back and forth) between the 5' nucleotide in the siRNA and a conserved tyrosine residue. This site is thought to form a nucleation site for the binding of the siRNA to its mRNA target. Analysis of the inhibitory effect of mismatches in either the 5' or 3' end of the guide strand has demonstrated that the 5' end of the guide strand is likely responsible for matching and binding the target mRNA, while the 3' end is responsible for physically arranging target mRNA into a cleavage-favorable RISC region.

It is not understood how the activated RISC complex locates complementary mRNAs within the cell. Although the cleavage process has been proposed to be linked to translation, translation of the mRNA target is not essential for RNAi-mediated degradation. Indeed, RNAi may be more effective against mRNA targets that are not translated. Argonaute proteins are localized to specific regions in the cytoplasm called P-bodies (also cytoplasmic bodies or GW bodies), which are regions with high rates of mRNA decay; miRNA activity is also clustered in P-bodies. Disruption of P-bodies decreases the efficiency of RNA interference, suggesting that they are a critical site in the RNAi process.

Transcriptional Silencing

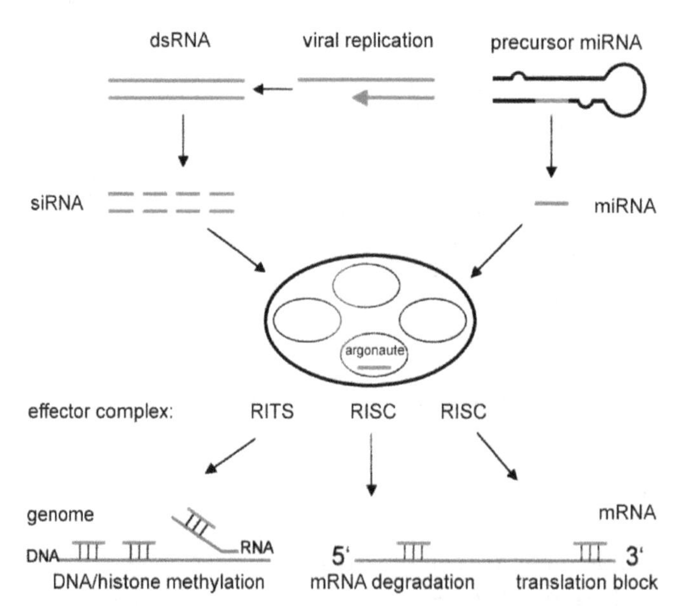

The enzyme dicer trims double stranded RNA, to form small interfering RNA or microRNA. These processed RNAs are incorporated into the RNA-induced silencing complex (RISC), which targets messenger RNA to prevent translation.

Components of the RNAi pathway are used in many eukaryotes in the maintenance of the organization and structure of their genomes. Modification of histones and associated induction of heterochromatin formation serves to downregulate genes pre-transcriptionally; this process is referred to as RNA-induced transcriptional silencing (RITS), and is carried out by a complex of proteins called the RITS complex. In fission yeast this complex contains argonaute, a chromodomain protein Chp1, and a protein called Tas3 of unknown function. As a consequence, the induction and spread of heterochromatic regions requires the argonaute and RdRP proteins. Indeed, deletion of these genes in the fission yeast *S. pombe* disrupts histone methylation and centromere formation, causing slow or stalled anaphase during cell division. In some cases, similar processes associated with histone modification have been observed to transcriptionally upregulate genes.

The mechanism by which the RITS complex induces heterochromatin formation and organization is not well understood. Most studies have focused on the mating-type region in fission yeast, which may not be representative of activities in other genomic regions/organisms. In maintenance of existing heterochromatin regions, RITS forms a complex with siRNAs complementary to the local genes and stably binds local methylated histones, acting co-transcriptionally to degrade any nascent pre-mRNA transcripts that are initiated by RNA polymerase. The formation of such a heterochromatin region, though not its maintenance, is dicer-dependent, presumably because dicer is required to generate the initial complement of siRNAs that target subsequent transcripts. Heterochromatin maintenance has been suggested to function as a self-reinforcing feedback loop, as new siRNAs are formed from the occasional nascent transcripts by RdRP for incorporation into local RITS complexes. The relevance of observations from fission yeast mating-type regions and centromeres to mammals is not clear, as heterochromatin maintenance in mammalian cells may be independent of the components of the RNAi pathway.

Crosstalk with RNA Editing

The type of RNA editing that is most prevalent in higher eukaryotes converts adenosine nucleotides into inosine in dsRNAs via the enzyme adenosine deaminase (ADAR). It was originally proposed in 2000 that the RNAi and A→I RNA editing pathways might compete for a common dsRNA substrate. Some pre-miRNAs do undergo A→I RNA editing and this mechanism may regulate the processing and expression of mature miRNAs. Furthermore, at least one mammalian ADAR can sequester siRNAs from RNAi pathway components. Further support for this model comes from studies on ADAR-null *C. elegans* strains indicating that A→I RNA editing may counteract RNAi silencing of endogenous genes and transgenes.

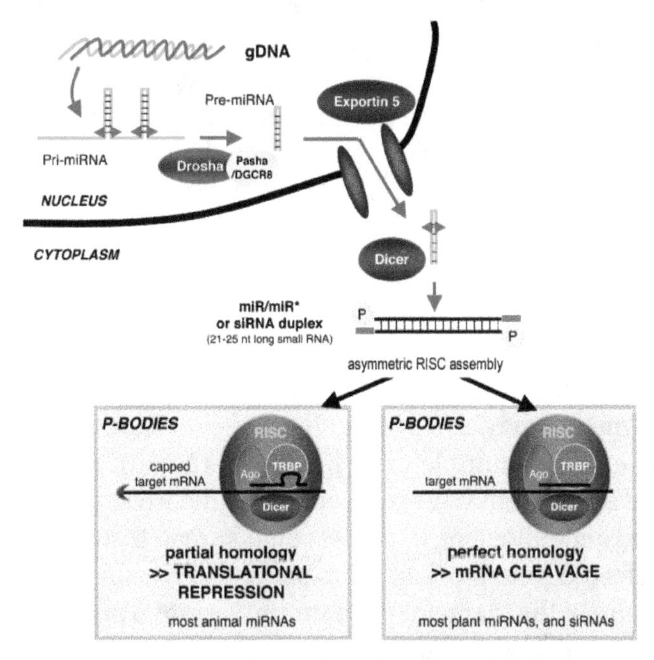

Illustration of the major differences between plant and animal gene silencing. Natively expressed microRNA or exogenous small interfering RNA is processed by dicer and integrated into the RISC complex, which mediates gene silencing.

Variation Among Organisms

Organisms vary in their ability to take up foreign dsRNA and use it in the RNAi pathway. The effects of RNA interference can be both systemic and heritable in plants and *C. elegans*, although not in *Drosophila* or mammals. In plants, RNAi is thought to propagate by the transfer of siRNAs between cells through plasmodesmata (channels in the cell walls that enable communication and transport). Heritability comes from methylation of promoters targeted by RNAi; the new methylation pattern is copied in each new generation of the cell. A broad general distinction between plants and animals lies in the targeting of endogenously produced miRNAs; in plants, miRNAs are usually perfectly or nearly perfectly complementary to their target genes and induce direct mRNA cleavage by RISC, while animals' miRNAs tend to be more divergent in sequence and induce translational repression. This translational effect may be produced by inhibiting the interactions of translation initiation factors with the messenger RNA's polyadenine tail.

Some eukaryotic protozoa such as *Leishmania major* and *Trypanosoma cruzi* lack the RNAi pathway entirely. Most or all of the components are also missing in some fungi, most notably the model organism *Saccharomyces cerevisiae*. The presence of RNAi in other budding yeast species such as *Saccharomyces castellii* and *Candida albicans*, further demonstrates that inducing two RNAi-related proteins from *S. castellii* facilitates RNAi in *S. cerevisiae*. That certain ascomycetes and basidiomycetes are missing RNA interference pathways indicates that proteins required for RNA silencing have been lost independently from many fungal lineages, possibly due to the evolution of a novel pathway with similar function, or to the lack of selective advantage in certain niches.

Related Prokaryotic Systems

Gene expression in prokaryotes is influenced by an RNA-based system similar in some respects to RNAi. Here, RNA-encoding genes control mRNA abundance or translation by producing a complementary RNA that anneals to an mRNA. However these regulatory RNAs are not generally considered to be analogous to miRNAs because the dicer enzyme is not involved. It has been suggested that CRISPR interference systems in prokaryotes are analogous to eukaryotic RNA interference systems, although none of the protein components are orthologous.

Biological Functions

Immunity

RNA interference is a vital part of the immune response to viruses and other foreign genetic material, especially in plants where it may also prevent the self-propagation of transposons. Plants such as *Arabidopsis thaliana* express multiple dicer homologs that are specialized to react differently when the plant is exposed to different viruses. Even before the RNAi pathway was fully understood, it was known that induced gene silencing in plants could spread throughout the plant in a systemic effect and could be transferred from stock to scion plants via grafting. This phenomenon has since been recognized as a feature of the plant adaptive immune system and allows the entire plant to respond to a virus after an initial localized encounter. In response, many plant viruses have evolved elaborate mechanisms to suppress the RNAi response. These include viral proteins that bind short double-stranded RNA fragments with single-stranded overhang ends, such as those produced by dicer. Some plant genomes also express endogenous siRNAs in response to infection by specific types of bacteria. These effects may be part of a generalized response to pathogens that downregulates any metabolic process in the host that aids the infection process.

Although animals generally express fewer variants of the dicer enzyme than plants, RNAi in some animals produces an antiviral response. In both juvenile and adult *Drosophila*, RNA interference is important in antiviral innate immunity and is active against pathogens such as Drosophila X virus. A similar role in immunity may operate in *C. elegans*, as argonaute proteins are upregulated in response to viruses and worms that overexpress components of the RNAi pathway are resistant to viral infection.

The role of RNA interference in mammalian innate immunity is poorly understood, and relatively little data is available. However, the existence of viruses that encode genes able to suppress the RNAi response in mammalian cells may be evidence in favour of an RNAi-dependent mammalian immune response, although this hypothesis has been challenged as poorly substantiated. Maillard et al. and Li et al. provide evidence for the existence of a functional antiviral RNAi pathway in mammalian cells. Other functions for RNAi in mammalian viruses also exist, such as miRNAs expressed by the herpes virus that may act as heterochromatin organization triggers to mediate viral latency.

Downregulation of Genes

Endogenously expressed miRNAs, including both intronic and intergenic miRNAs, are most important in translational repression and in the regulation of development, especially on the timing of morphogenesis and the maintenance of undifferentiated or incompletely differentiated cell types such as stem cells. The role of endogenously expressed miRNA in downregulating gene expression was first described in *C. elegans* in 1993. In plants this function was discovered when the "JAW microRNA" of *Arabidopsis* was shown to be involved in the regulation of several genes that control plant shape. In plants, the majority of genes regulated by miRNAs are transcription factors; thus miRNA activity is particularly wide-ranging and regulates entire gene networks during development by modulating the expression of key regulatory genes, including transcription factors as well as F-box proteins. In many organisms, including humans, miRNAs are linked to the formation of tumors and dysregulation of the cell cycle. Here, miRNAs can function as both oncogenes and tumor suppressors.

Upregulation of Genes

RNA sequences (siRNA and miRNA) that are complementary to parts of a promoter can increase gene transcription, a phenomenon dubbed RNA activation. Part of the mechanism for how these RNA upregulate genes is known: dicer and argonaute are involved, possibly via histone demethylation. miRNAs have been proposed to upregulate their target genes upon cell cycle arrest, via unknown mechanisms.

Evolution

Based on parsimony-based phylogenetic analysis, the most recent common ancestor of all eukaryotes most likely already possessed an early RNA interference pathway; the absence of the pathway in certain eukaryotes is thought to be a derived characteristic. This ancestral RNAi system probably contained at least one dicer-like protein, one argonaute, one PIWI protein, and an RNA-dependent RNA polymerase that may also have played other cellular roles. A large-scale comparative genomics study likewise indicates that the eukaryotic crown group already possessed these components, which may then have had closer functional associations with generalized RNA degradation systems such as the exosome. This study also suggests that the RNA-binding argonaute protein family, which is shared among eukaryotes, most archaea, and at least some bacteria (such as *Aquifex aeolicus*), is homologous to and originally evolved from components of the translation initiation system.

The ancestral function of the RNAi system is generally agreed to have been immune defense against exogenous genetic elements such as transposons and viral genomes. Related functions such as histone modification may have already been present in the ancestor of modern eukaryotes, although other functions such as regulation of development by miRNA are thought to have evolved later.

RNA interference genes, as components of the antiviral innate immune system in many eukaryotes, are involved in an evolutionary arms race with viral genes. Some viruses have evolved mechanisms for suppressing the RNAi response in their host cells, particularly for plant viruses. Studies of evolutionary rates in *Drosophila* have shown that genes in the RNAi pathway are subject to strong directional selection and are among the fastest-evolving genes in the *Drosophila* genome.

Applications

Gene Knockdown

The RNA interference pathway is often exploited in experimental biology to study the function of genes in cell culture and *in vivo* in model organisms. Double-stranded RNA is synthesized with a sequence complementary to a gene of interest and introduced into a cell or organism, where it is recognized as exogenous genetic material and activates the RNAi pathway. Using this mechanism, researchers can cause a drastic decrease in the expression of a targeted gene. Studying the effects of this decrease can show the physiological role of the gene product. Since RNAi may not totally abolish expression of the gene, this technique is sometimes referred as a "knockdown", to distinguish it from "knockout" procedures in which expression of a gene is entirely eliminated.

Extensive efforts in computational biology have been directed toward the design of successful dsRNA reagents that maximize gene knockdown but minimize "off-target" effects. Off-target effects arise when an introduced RNA has a base sequence that can pair with and thus reduce the expression of multiple genes. Such problems occur more frequently when the dsRNA contains repetitive sequences. It has been estimated from studying the genomes of humans, *C. elegans* and *S. pombe* that about 10% of possible siRNAs have substantial off-target effects. A multitude of software tools have been developed implementing algorithms for the design of general mammal-specific, and virus-specific siRNAs that are automatically checked for possible cross-reactivity.

Depending on the organism and experimental system, the exogenous RNA may be a long strand designed to be cleaved by dicer, or short RNAs designed to serve as siRNA substrates. In most mammalian cells, shorter RNAs are used because long double-stranded RNA molecules induce the mammalian interferon response, a form of innate immunity that reacts nonspecifically to foreign genetic material. Mouse oocytes and cells from early mouse embryos lack this reaction to exogenous dsRNA and are therefore a common model system for studying mammalian gene-knockdown effects. Specialized laboratory techniques have also been developed to improve the utility of RNAi in mammalian systems by avoiding the direct introduction of siRNA, for example, by stable transfection with a plasmid encoding the appropriate sequence from which siRNAs can be transcribed, or by more elaborate lentiviral vector systems allowing the inducible activation or deactivation of transcription, known as *conditional RNAi.*

Functional Genomics

Most functional genomics applications of RNAi in animals have used *C. elegans* and *Drosophila,* as these are the common model organisms in which RNAi is most effective. *C. elegans* is particularly useful for RNAi research for two reasons: firstly, the effects of gene silencing are generally heritable, and secondly because delivery of the dsRNA is extremely simple. Through a mechanism whose details are poorly understood, bacteria such as *E. coli* that carry the desired dsRNA can be fed to the worms and will transfer their RNA payload to the worm via the intestinal tract. This "delivery by feeding" is just as effective at inducing gene silencing as more costly and time-consuming delivery methods, such as soaking the worms in dsRNA solution and injecting dsRNA into the gonads. Although delivery is more difficult in most other organisms, efforts are also underway to undertake large-scale genomic screening applications in cell culture with mammalian cells.

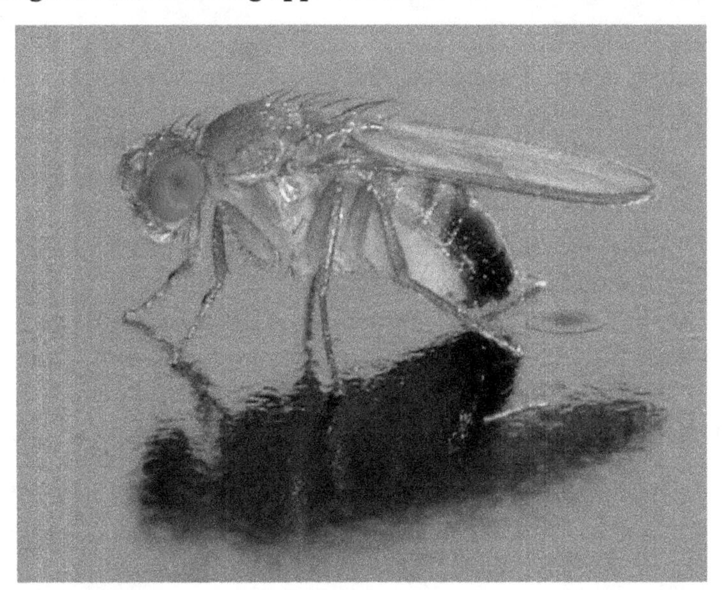

A normal adult *Drosophila* fly, a common model organism used in RNAi experiments.

Approaches to the design of genome-wide RNAi libraries can require more sophistication than the design of a single siRNA for a defined set of experimental conditions. Artificial neural networks are frequently used to design siRNA libraries and to predict their likely efficiency at gene knockdown. Mass genomic screening is widely seen as a promising method for genome annotation and has triggered the development of high-throughput screening methods based on microarrays. However, the utility of these screens and the ability of techniques developed on model organisms to generalize to even closely related species has been questioned, for example from *C. elegans* to related parasitic nematodes.

Functional genomics using RNAi is a particularly attractive technique for genomic mapping and annotation in plants because many plants are polyploid, which presents substantial challenges for more traditional genetic engineering methods. For example, RNAi has been successfully used for functional genomics studies in bread wheat (which is hexaploid) as well as more common plant model systems *Arabidopsis* and maize.

Medicine

It may be possible to exploit RNA interference in therapy. Although it is difficult to introduce long dsRNA strands into mammalian cells due to the interferon response, the use of short interfering RNA has been more successful. Among the first applications to reach clinical trials were in the treatment of macular degeneration and respiratory syncytial virus. RNAi has also been shown to be effective in reversing induced liver failure in mouse models.

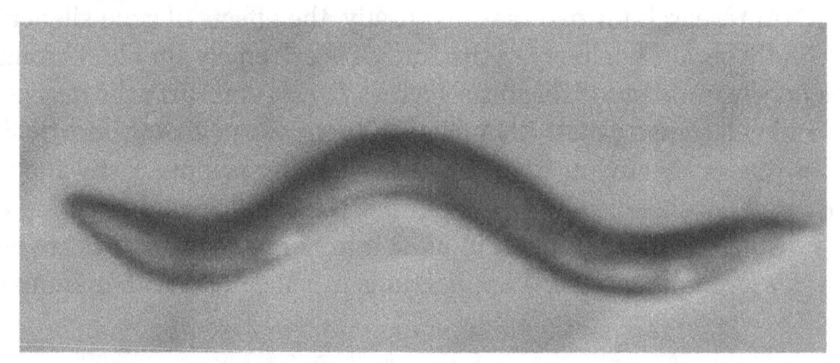

An adult *C. elegans* worm, grown under RNAi suppression of a nuclear hormone receptor involved in desaturase regulation. These worms have abnormal fatty acid metabolism but are viable and fertile.

Antiviral

Potential antiviral therapies include topical microbicide treatments that use RNAi to treat infection (at Harvard Medical School; in mice, so far) by herpes simplex virus type 2 and the inhibition of viral gene expression in cancerous cells, knockdown of host receptors and coreceptors for HIV, the silencing of hepatitis A and hepatitis B genes, silencing of influenza gene expression, and inhibition of measles viral replication. Potential treatments for neurodegenerative diseases have also been proposed, with particular attention to polyglutamine diseases such as Huntington's disease.

RNA interference-based applications are being developed to target persistent HIV-1 infection. Viruses like HIV-1 are particularly difficult targets for RNAi-attack because they are escape-prone, which requires combinatorial RNAi strategies to prevent viral escape.

Cancer

RNA interference is also a promising way to treat cancers by silencing genes differentially upregulated in tumor cells or genes involved in cell division. A key area of research in the use of RNAi for clinical applications is the development of a safe delivery method, which to date has involved mainly viral vector systems similar to those suggested for gene therapy.

Due to safety concerns with viral vectors, nonviral delivery methods, typically employing lipid-based or polymeric vectors, are also promising candidates. Computational modeling of nonviral siRNA delivery paired with *in vitro* and *in vivo* gene knockdown studies elucidated the temporal behavior of RNAi in these systems. The model used an input bolus dose of siRNA and computationally and experimentally showed that knockdown duration was dependent mainly on the doubling time of the cells to which siRNA was delivered, while peak knockdown depended primarily on the delivered dose. Kinetic considerations of RNAi are imperative to safe and effective dosing schedules as nonviral methods of inducing RNAi continue to be developed.

Safety

Despite the proliferation of promising cell culture studies for RNAi-based drugs, some concern has been raised regarding the safety of RNA interference, especially the potential for "off-target" effects in which a gene with a coincidentally similar sequence to the targeted gene is also repressed. A computational genomics study estimated that the error rate of off-target interactions is about 10%. One major study of liver disease in mice reported that 23 out of 49 distinct RNAi treatment protocols resulted in death. Researchers hypothesized this alarmingly high rate to be the result of "oversaturation" of the dsRNA pathway, due to the use of shRNAs that have to be processed in the nucleus and exported to the cytoplasm using an active mechanism. Such considerations are under active investigation, to reduce their impact in the potential therapeutic applications.

RNAi in vivo delivery to tissues still eludes science—especially to tissues deep within the body. RNAi delivery is only easily accessible to surface tissues such as the eye and respiratory tract. In these instances, siRNA has been used in direct contact with the tissue for transport. The resulting RNAi successfully focused on target genes. When delivering siRNA to deep tissues, the siRNA must be protected from nucleases, but targeting specific areas becomes the main difficulty. This difficulty has been combatted with high dosage levels of siRNA to ensure the tissues have been reached, however in these cases hepatotoxicity was reported.

Biotechnology

RNA interference has been used for applications in biotechnology and is nearing commercialization in others. RNAi has developed many novel crops such as nicotinefree tobacco, decaffeinated coffee, nutrient fortified and hypoallergenic crops. The genetically engineered Arctic apples are near close to receive US approval. The apples were produced by RNAi suppression of PPO (polyphenol oxidase) gene making apple varieties that will not undergo browning after being sliced. PPO-silenced apples are unable to convert chlorogenic acid into quinone product.

There are several opportunities for the applications of RNAi in crop science for its improvement such as stress tolerance and enhanced nutritional level. RNAi will prove its potential for inhibition of photorespiration to enhance the productivity of C3 plants. This knockdown technology may be useful in inducing early flowering, delayed ripening, delayed senescence, breaking dormancy, stress-free plants, overcoming self-sterility, etc.

Foods

RNAi has been used to genetically engineer plants to produce lower levels of natural plant toxins. Such techniques take advantage of the stable and heritable RNAi phenotype in plant stocks. Cotton seeds are rich in dietary protein but naturally contain the toxic terpenoid product gossypol, making them unsuitable for human consumption. RNAi has been used to produce cotton stocks whose seeds contain reduced levels of delta-cadinene synthase, a key enzyme in gossypol production, without affecting the enzyme's production in other parts of the plant, where gossypol is itself important in preventing damage from plant pests. Similar efforts have been directed toward the reduction of the cyanogenic natural product linamarin in cassava plants.

No plant products that use RNAi-based genetic engineering have yet exited the experimental stage. Development efforts have successfully reduced the levels of allergens in tomato plants and fortification of plants such as tomatoes with dietary antioxidants. Previous commercial products, including the Flavr Savr tomato and two cultivars of ringspot-resistant papaya, were originally developed using antisense technology but likely exploited the RNAi pathway.

Other Crops

Another effort decreased the precursors of likely carcinogens in tobacco plants. Other plant traits that have been engineered in the laboratory include the production of non-narcotic natural products by the opium poppy and resistance to common plant viruses.

Insecticide

RNAi is under development as an insecticide, employing multiple approaches, including genetic engineering and topical application. Cells in the midgut of many larvae take up the molecules and help spread the signal throughout the insect's body.

RNAi has varying effects in different species of Lepidoptera (butterflies and moths). Possibly because their saliva is better at breaking down RNA, the cotton bollworm, the beet armyworm and the Asiatic rice borer have so far not been proven susceptible to RNAi by feeding.

To develop resistance to RNAi, the western corn rootworm would have to change the genetic sequence of its Snf7 gene at multiple sites. Combining multiple strategies, such as engineering the protein Cry, derived from a bacterium called Bacillus thuringiensis (Bt), and RNAi in one plant delay the onset of resistance.

Transgenic Plants

Transgenic crops have been made to express small bits of RNA, carefully chosen to silence crucial genes in target pests. RNAs exist that affect only insects that have specific genetic sequences. In 2009 a study showed RNAs that could kill any one of four fruit fly species while not harming the other three.

In 2012 Syngenta bought Belgian RNAi firm Devgen for $522 million and Monsanto paid $29.2 million for the exclusive rights to intellectual property from Alnylam Pharmaceuticals. The International Potato Center in Lima, Peru is looking for genes to target in the sweet potato weevil, a beetle whose larvae ravage sweet potatoes globally. Other researchers are trying to silence genes in ants, caterpillars and pollen beetles. Monsanto will likely be first to market, with a transgenic corn seed that expresses dsRNA based on gene Snf7 from the western corn rootworm, a beetle whose larvae annually cause one billion dollars in damage in the United States alone. A 2012 paper showed that silencing Snf7 stunts larval growth, killing them within days. In 2013 the same team showed that the RNA affects very few other species.

Topical

Alternatively dsRNA can be supplied without genetic engineering. One approach is to add them to irrigation water. The molecules are absorbed into the plants' vascular system and poison insects feeding on them. Another approach involves spraying RNA like a conventional pesticide. This would allow faster adaptation to resistance. Such approaches would require low cost sources of RNAs that do not currently exist.

Genome-scale Screening

Genome-scale RNAi research relies on high-throughput screening (HTS) technology. RNAi HTS technology allows genome-wide loss-of-function screening and is broadly used in the identification of genes associated with specific phenotypes. This technology has been hailed as the second genomics wave, following the first genomics wave of gene expression microarray and single nucleotide polymorphism discovery platforms. One major advantage of genome-scale RNAi screening is its ability to simultaneously interrogate thousands of genes. With the ability to generate a large amount of data per experiment, genome-scale RNAi screening has led to an explosion data generation rates. Exploiting such large data sets is a fundamental challenge, requiring suitable statistics/ bioinformatics methods. The basic process of cell-based RNAi screening includes the choice of an RNAi library, robust and stable cell types, transfection with RNAi agents, treatment/incubation, signal detection, analysis and identification of important genes or therapeutical targets.

History

Example petunia plants in which genes for pigmentation are silenced by RNAi. The left plant is wild-type; the right plants contain transgenes that induce suppression of both transgene and endogenous gene expression, giving rise to the unpigmented white areas of the flower.

The discovery of RNAi was preceded first by observations of transcriptional inhibition by antisense RNA expressed in transgenic plants, and more directly by reports of unexpected outcomes in experiments performed by plant scientists in the United States and the Netherlands in the early 1990s. In an attempt to alter flower colors in petunias, researchers introduced additional copies of a gene encoding chalcone synthase, a key enzyme for flower pigmentation into petunia plants of normally pink or violet flower color. The overexpressed gene was expected to result in darker flowers, but instead produced less pigmented, fully or partially white flowers, indicating that the activity of chalcone synthase had been substantially decreased; in fact, both the endogenous genes and the transgenes were downregulated in the white flowers. Soon after, a related event termed *quelling* was noted in the fungus *Neurospora crassa*, although it was not immediately recognized as related. Further investigation of the phenomenon in plants indicated that the downregulation was due to post-transcriptional inhibition of gene expression via an increased rate of mRNA degradation. This phenomenon was called *co-suppression of gene expression*, but the molecular mechanism remained unknown.

Not long after, plant virologists working on improving plant resistance to viral diseases observed a similar unexpected phenomenon. While it was known that plants expressing virus-specific proteins showed enhanced tolerance or resistance to viral infection, it was not expected that plants carrying only short, non-coding regions of viral RNA sequences would show similar levels of protection. Researchers believed that viral RNA produced by transgenes could also inhibit viral replication. The reverse experiment, in which short sequences of plant genes were introduced into viruses, showed that the targeted gene was suppressed in an infected plant. This phenomenon was labeled "virus-induced gene silencing" (VIGS), and the set of such phenomena were collectively called *post transcriptional gene silencing*.

After these initial observations in plants, laboratories searched for this phenomenon in other organisms. Craig C. Mello and Andrew Fire's 1998 *Nature* paper reported a potent gene silencing effect after injecting double stranded RNA into *C. elegans*. In investigating the regulation of muscle protein production, they observed that neither mRNA nor antisense RNA injections had an effect on protein production, but double-stranded RNA successfully silenced the targeted gene. As a result of this work, they coined the term *RNAi*. This discovery represented the first identification of the causative agent for the phenomenon. Fire and Mello were awarded the 2006 Nobel Prize in Physiology or Medicine.

References

- Imperfect DNA replication results in mutations. Berg JM, Tymoczko JL, Stryer L, Clarke ND (2002). "Chapter 27: DNA Replication, Recombination, and Repair". Biochemistry. W.H. Freeman and Company. ISBN 0-7167-3051-0. External link in |chapter= (help)
- Alberts B, Johnson A, Lewis J, Raff M, Roberts K, Walter P (2002). "5DNA Replication, Repair, and Recombination". Molecular Biology of the Cell. Garland Science. ISBN 0-8153-3218-1. External link in |chapter= (help)
- Berg JM, Tymoczko JL, Stryer L, Clarke ND (2002). "Chapter 27, Section 4: DNA Replication of Both Strands Proceeds Rapidly from Specific Start Sites". Biochemistry. W.H. Freeman and Company. ISBN 0-7167-3051-0.
- Lodish H, Berk A, Zipursky LS, Matsudaira P, Baltimore D, Darnell J (2000). Molecular Cell Biology. W. H. Freeman and Company. ISBN 0-7167-3136-3.12.1.
- Alberts B, Johnson A, Lewis J, Raff M, Roberts K, Walter P (2002). Molecular Biology of the Cell. Garland Science. ISBN 0-8153-3218-1.
- Alberts B, Johnson A, Lewis J, Raff M, Roberts K, Walter P (2002). Molecular Biology of the Cell. Garland Science. ISBN 0-8153-3218-1.
- Griffiths A.J.F.; Wessler S.R.; Lewontin R.C.; Carroll S.B. (2008). Introduction to Genetic Analysis. W. H. Freeman and Company. ISBN 0-7167-6887-9.
- Alberts B, Johnson A, Lewis J, Raff M, Roberts K, Walter P (2002). Molecular Biology of the Cell. Garland Science. ISBN 0-8153-3218-1.
- Eldra P. Solomon, Linda R. Berg, Diana W. Martin. Biology, 8th Edition, International Student Edition. Thomson Brooks/Cole. ISBN 978-0495317142
- Alberts, Bruce; Alexander Johnson; Julian Lewis; Martin Raff; Keith Roberts; Peter Walters (2002). "The Shape and Structure of Proteins". Molecular Biology of the Cell; Fourth Edition. New York and London: Garland Science. ISBN 0-8153-3218-1.
- Jeremy M. Berg, John L. Tymoczko, Lubert Stryer; Web content by Neil D. Clarke (2002). "3. Protein Structure and Function". Biochemistry. San Francisco: W. H. Freeman. ISBN 0-7167-4684-0.
- Neill, Campbell (1996). Biology; Fourth edition. The Benjamin/Cummings Publishing Company. p. 309,310. ISBN 0-8053-1940-9.
- Griffiths, Anthony (2008). "9". Introduction to Genetic Analysis (9th ed.). New York: W.H. Freeman and Com-

pany. pp. 335–339. ISBN 978-0-7167-6887-6.

- Jones, Daniel (2003) [1917], Peter Roach, James Hartmann and Jane Setter, eds., English Pronouncing Dictionary, Cambridge: Cambridge University Press, ISBN 3-12-539683-2
- Gregory R, Chendrimada T, Shiekhattar R (2006). "MicroRNA biogenesis: isolation and characterization of the microprocessor complex". Methods Mol Biol. 342: 33–47. doi:10.1385/1-59745-123-1:33. ISBN 1-59745-123-1.
- Lodish H, Berk A, Matsudaira P, Kaiser CA, Krieger M, Scott MP, Zipurksy SL, Darnell J (2004). Molecular Cell Biology (5th ed.). WH Freeman: New York, NY. ISBN 978-0-7167-4366-8.
- Janitz, M; Vanhecke, D; Lehrach, H (2006). "High-throughput RNA interference in functional genomics". Handb Exp Pharmacol. Handbook of Experimental Pharmacology. 173 (173): 97–104. doi:10.1007/3-540-27262-3_5. ISBN 3-540-27261-5.
- Zhang XHD (2011). Optimal High-Throughput Screening: Practical Experimental Design and Data Analysis for Genome-scale RNAi Research. Cambridge University Press. ISBN 978-0-521-73444-8.
- Keeley, Jim; Gutnikoff, Robert (2015-01-02). "Ribosome Studies Turn Up New Mechanism of Protein Synthesis" (Press release). Howard Hughes Medical Institute. Retrieved 2015-01-16.

3

Techniques of Molecular Biology

Molecular cloning is a method used to assemble recombinant DNA molecules and polymerase chain reaction is the copying of a piece of DNA to generate millions of copies of the same DNA. This chapter discusses the techniques of molecular biology in a critical manner providing key analysis to the subject matter.

Molecular Cloning

Diagram of molecular cloning using bacteria and plasmids.

Molecular cloning is a set of experimental methods in molecular biology that are used to assemble recombinant DNA molecules and to direct their replication within host organisms. The use of the word *cloning* refers to the fact that the method involves the replication of one molecule to produce a population of cells with identical DNA molecules. Molecular cloning generally uses DNA sequences from two different organisms: the species that is the source of the DNA to be cloned, and the species that will serve as the living host for replication of the recombinant DNA. Molecular cloning methods are central to many contemporary areas of modern biology and medicine.

In a conventional molecular cloning experiment, the DNA to be cloned is obtained from an organism of interest, then treated with enzymes in the test tube to generate smaller DNA fragments. Subsequently, these fragments are then combined with vector DNA to generate recombinant DNA molecules. The recombinant DNA is then introduced into a host organism (typically an easy-to-grow, benign, laboratory strain of *E. coli* bacteria). This will generate a population of organisms in which recombinant DNA molecules are replicated along with the host DNA. Because they contain foreign DNA fragments, these are transgenic or genetically modified microorganisms (GMO). This process takes advantage of the fact that a single bacterial cell can be induced to take up and replicate a single recombinant DNA molecule. This single cell can then be expanded exponentially to generate a large amount of bacteria, each of which contain copies of the original recombinant molecule. Thus, both the resulting bacterial population, and the recombinant DNA molecule, are commonly referred to as "clones". Strictly speaking, *recombinant DNA* refers to DNA molecules, while *molecular cloning* refers to the experimental methods used to assemble them.

History of Molecular Cloning

Prior to the 1970s, our understanding of genetics and molecular biology was severely hampered by an inability to isolate and study individual genes from complex organisms. This changed dramatically with the advent of molecular cloning methods. Microbiologists, seeking to understand the molecular mechanisms through which bacteria restricted the growth of bacteriophage, isolated restriction endonucleases, enzymes that could cleave DNA molecules only when specific DNA sequences were encountered. They showed that restriction enzymes cleaved chromosome-length DNA molecules at specific locations, and that specific sections of the larger molecule could be purified by size fractionation. Using a second enzyme, DNA ligase, fragments generated by restriction enzymes could be joined in new combinations, termed recombinant DNA. By recombining DNA segments of interest with vector DNA, such as bacteriophage or plasmids, which naturally replicate inside bacteria, large quantities of purified recombinant DNA molecules could be produced in bacterial cultures. The first recombinant DNA molecules were generated and studied in 1972.

Overview

Molecular cloning takes advantage of the fact that the chemical structure of DNA is fundamentally the same in all living organisms. Therefore, if any segment of DNA from any organism is inserted into a DNA segment containing the molecular sequences required for DNA replication, and the resulting recombinant DNA is introduced into the organism from which the replication sequences were obtained, then the foreign DNA will be replicated along with the host cell's DNA in the transgenic organism.

Molecular cloning is similar to polymerase chain reaction (PCR) in that it permits the replication of DNA sequence. The fundamental difference between the two methods is that molecular cloning involves replication of the DNA in a living microorganism, while PCR replicates DNA in an *in vitro* solution, free of living cells.

Steps in Molecular Cloning

In standard molecular cloning experiments, the cloning of any DNA fragment essentially involves seven steps: (1) Choice of host organism and cloning vector, (2) Preparation of vector DNA, (3) Preparation of DNA to be cloned, (4) Creation of recombinant DNA, (5) Introduction of recombinant DNA into host organism, (6) Selection of organisms containing recombinant DNA, (7) Screening for clones with desired DNA inserts and biological properties.

Although the detailed planning of the cloning can be done in any text editor, together with online utilities for e.g. PCR primer design, dedicated software exist for the purpose. Software for the purpose include for example ApE (open source), DNAStrider (open source), Serial Cloner (gratis) and Collagene (open source).

Choice of Host Organism and Cloning Vector

Although a very large number of host organisms and molecular cloning vectors are in use, the great majority of molecular cloning experiments begin with a laboratory strain of the bacterium *E. coli* (*Escherichia coli*) and a plasmid cloning vector. *E. coli* and plasmid vectors are in common use because they are technically sophisticated, versatile, widely available, and offer rapid growth of recombinant organisms with minimal equipment. If the DNA to be cloned is exceptionally large (hundreds of thousands to millions of base pairs), then a bacterial artificial chromosome or yeast artificial chromosome vector is often chosen.

Specialized applications may call for specialized host-vector systems. For example, if the experimentalists wish to harvest a particular protein from the recombinant organism, then an expression vector is chosen that contains appropriate signals for transcription and translation in the desired host organism. Alternatively, if replication of the DNA in different species is desired (for example, transfer of DNA from bacteria to plants), then a multiple host range vector (also termed shuttle vector) may be selected. In practice, however, specialized molecular cloning experiments usually begin with cloning into a bacterial plasmid, followed by subcloning into a specialized vector.

Diagram of a commonly used cloning plasmid; pBR322. It's a circular piece of DNA 4361 bases long. Two antibiotic resistance genes are present, conferring resistance to ampicillin and tetracycline, and an origin of replication that the host uses to replicate the DNA.

Whatever combination of host and vector are used, the vector almost always contains four DNA segments that are critically important to its function and experimental utility:

1. DNA *replication origin* is necessary for the vector (and its linked recombinant sequences) to replicate inside the host organism
2. one or more unique *restriction endonuclease recognition sites* to serves as sites where foreign DNA may be introduced
3. a *selectable genetic marker* gene that can be used to enable the survival of cells that have taken up vector sequences
4. a *tag* gene that can be used to screen for cells containing the foreign DNA

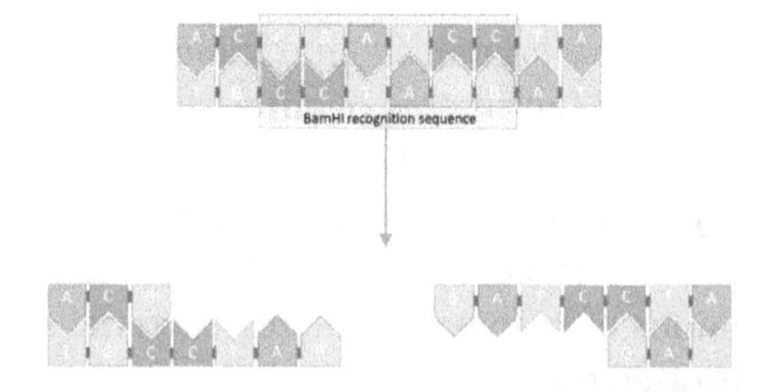

Cleavage of a DNA sequence containing the BamHI restriction site. The DNA is cleaved at the palindromic sequence to produce 'sticky ends'.

Preparation of Vector DNA

The cloning vector is treated with a restriction endonuclease to cleave the DNA at the site where foreign DNA will be inserted. The restriction enzyme is chosen to generate a configuration at the cleavage site that is compatible with the ends of the foreign DNA. Typically, this is done by cleaving the vector DNA and foreign DNA with the same restriction enzyme, for example EcoRI. Most modern vectors contain a variety of convenient cleavage sites that are unique within the vector molecule (so that the vector can only be cleaved at a single site) and are located within a gene (frequently beta-galactosidase) whose inactivation can be used to distinguish recombinant from non-recombinant organisms at a later step in the process. To improve the ratio of recombinant to non-recombinant organisms, the cleaved vector may be treated with an enzyme (alkaline phosphatase) that dephosphorylates the vector ends. Vector molecules with dephosphorylated ends are unable to replicate, and replication can only be restored if foreign DNA is integrated into the cleavage site.

Preparation of DNA to be Cloned

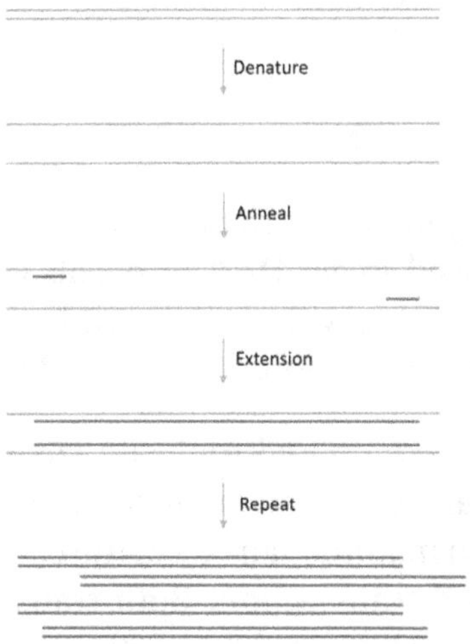

Denature

Anneal

Extension

Repeat

DNA for cloning is most commonly produced using PCR. Template DNA is mixed with bases (the building blocks of DNA), primers (short pieces of complementary single stranded DNA) and a DNA polymerase enzyme that builds the DNA chain. The mix goes through cycles of heating and cooling to produce large quantities of copied DNA.

For cloning of genomic DNA, the DNA to be cloned is extracted from the organism of interest. Virtually any tissue source can be used (even tissues from extinct animals), as long as the DNA is not extensively degraded. The DNA is then purified using simple methods to remove contaminating proteins (extraction with phenol), RNA (ribonuclease) and smaller molecules (precipitation and/or chromatography). Polymerase chain reaction (PCR) methods are often used for amplification of specific DNA or RNA (RT-PCR) sequences prior to molecular cloning.

DNA for cloning experiments may also be obtained from RNA using reverse transcriptase (complementary DNA or cDNA cloning), or in the form of synthetic DNA (artificial gene synthesis). cDNA cloning is usually used to obtain clones representative of the mRNA population of the cells of interest, while synthetic DNA is used to obtain any precise sequence defined by the designer.

The purified DNA is then treated with a restriction enzyme to generate fragments with ends capable of being linked to those of the vector. If necessary, short double-stranded segments of DNA (*linkers*) containing desired restriction sites may be added to create end structures that are compatible with the vector.

Creation of Recombinant DNA with DNA Ligase

The creation of recombinant DNA is in many ways the simplest step of the molecular cloning process. DNA prepared from the vector and foreign source are simply mixed together at appropriate concentrations and exposed to an enzyme (DNA ligase) that covalently links the ends together. This joining reaction is often termed ligation. The resulting DNA mixture containing randomly joined ends is then ready for introduction into the host organism.

DNA ligase only recognizes and acts on the ends of linear DNA molecules, usually resulting in a complex mixture of DNA molecules with randomly joined ends. The desired products (vector DNA covalently linked to foreign DNA) will be present, but other sequences (e.g. foreign DNA linked to itself, vector DNA linked to itself and higher-order combinations of vector and foreign DNA) are also usually present. This complex mixture is sorted out in subsequent steps of the cloning process, after the DNA mixture is introduced into cells.

Introduction of Recombinant DNA Into Host Organism

The DNA mixture, previously manipulated in vitro, is moved back into a living cell, referred to as the host organism. The methods used to get DNA into cells are varied, and the name applied to this step in the molecular cloning process will often depend upon the experimental method that is chosen (e.g. transformation, transduction, transfection, electroporation).

When microorganisms are able to take up and replicate DNA from their local environment, the process is termed transformation, and cells that are in a physiological state such that they can take up DNA are said to be competent. In mammalian cell culture, the analogous process of introducing DNA into cells is commonly termed transfection. Both transformation and transfection usually require preparation of the cells through a special growth regime and chemical treatment process that will vary with the specific species and cell types that are used.

Electroporation uses high voltage electrical pulses to translocate DNA across the cell membrane (and cell wall, if present). In contrast, transduction involves the packaging of DNA into virus-derived particles, and using these virus-like particles to introduce the encapsulated DNA into the cell through a process resembling viral infection. Although electroporation and transduction are highly specialized methods, they may be the most efficient methods to move DNA into cells.

Selection of Organisms Containing Vector Sequences

Whichever method is used, the introduction of recombinant DNA into the chosen host organism is usually a low efficiency process; that is, only a small fraction of the cells will actually take up DNA. Experimental scientists deal with this issue through a step of artificial genetic selection, in which cells that have not taken up DNA are selectively killed, and only those cells that can actively replicate DNA containing the selectable marker gene encoded by the vector are able to survive.

When bacterial cells are used as host organisms, the selectable marker is usually a gene that confers resistance to an antibiotic that would otherwise kill the cells, typically ampicillin. Cells harboring the plasmid will survive when exposed to the antibiotic, while those that have failed to take up plasmid sequences will die. When mammalian cells (e.g. human or mouse cells) are used, a similar strategy is used, except that the marker gene (in this case typically encoded as part of the kanMX cassette) confers resistance to the antibiotic Geneticin.

Screening for Clones with Desired DNA Inserts and Biological Properties

Modern bacterial cloning vectors (e.g. pUC19 and later derivatives including the pGEM vectors) use the blue-white screening system to distinguish colonies (clones) of transgenic cells from those that contain the parental vector (i.e. vector DNA with no recombinant sequence inserted). In these vectors, foreign DNA is inserted into a sequence that encodes an essential part of beta-galactosidase, an enzyme whose activity results in formation of a blue-colored colony on the culture medium that is used for this work. Insertion of the foreign DNA into the beta-galactosidase coding sequence disables the function of the enzyme, so that colonies containing transformed DNA remain colorless (white). Therefore, experimentalists are easily able to identify and conduct further studies on transgenic bacterial clones, while ignoring those that do not contain recombinant DNA.

The total population of individual clones obtained in a molecular cloning experiment is often termed a DNA library. Libraries may be highly complex (as when cloning complete genomic DNA from an organism) or relatively simple (as when moving a previously cloned DNA fragment into a different plasmid), but it is almost always necessary to examine a number of different clones to be sure that the desired DNA construct is obtained. This may be accomplished through a very wide range of experimental methods, including the use of nucleic acid hybridizations, antibody probes, polymerase chain reaction, restriction fragment analysis and/or DNA sequencing.

Applications of Molecular Cloning

Molecular cloning provides scientists with an essentially unlimited quantity of any individual DNA segments derived from any genome. This material can be used for a wide range of purposes, including those in both basic and applied biological science. A few of the more important applications are summarized here.

Genome Organization and Gene Expression

Molecular cloning has led directly to the elucidation of the complete DNA sequence of the genomes of a very large number of species and to an exploration of genetic diversity within individual species, work that has been done mostly by determining the DNA sequence of large numbers of randomly cloned fragments of the genome, and assembling the overlapping sequences.

At the level of individual genes, molecular clones are used to generate probes that are used for examining how genes are expressed, and how that expression is related to other processes in biology, including the metabolic environment, extracellular signals, development, learning, senescence and cell death. Cloned genes can also provide tools to examine the biological function and importance of individual genes, by allowing investigators to inactivate the genes, or make more subtle mutations using regional mutagenesis or site-directed mutagenesis.

Production of Recombinant Proteins

Obtaining the molecular clone of a gene can lead to the development of organisms that produce the protein product of the cloned genes, termed a recombinant protein. In practice, it is frequently more difficult to develop an organism that produces an active form of the recombinant protein in desirable quantities than it is to clone the gene. This is because the molecular signals for gene expression are complex and variable, and because protein folding, stability and transport can be very challenging.

Many useful proteins are currently available as recombinant products. These include-(1) medically useful proteins whose administration can correct a defective or poorly expressed gene (e.g. recombinant factor VIII, a blood-clotting factor deficient in some forms of hemophilia, and recombinant insulin, used to treat some forms of diabetes), (2) proteins that can be administered to assist in a life-threatening emergency (e.g. tissue plasminogen activator, used to treat strokes), (3) recombinant subunit vaccines, in which a purified protein can be used to immunize patients against infectious diseases, without exposing them to the infectious agent itself (e.g. hepatitis B vaccine), and (4) recombinant proteins as standard material for diagnostic laboratory tests.

Transgenic Organisms

Once characterized and manipulated to provide signals for appropriate expression, cloned genes may be inserted into organisms, generating transgenic organisms, also termed genetically modified organisms (GMOs). Although most GMOs are generated for purposes of basic biological research, a number of GMOs have been developed for commercial use, ranging from animals and plants that produce pharmaceuticals or other compounds (pharming), herbicide-resistant crop plants, and fluorescent tropical fish (GloFish) for home entertainment.

Gene Therapy

Gene therapy involves supplying a functional gene to cells lacking that function, with the aim of correcting a genetic disorder or acquired disease. Gene therapy can be broadly divided into two categories. The first is alteration of germ cells, that is, sperm or eggs, which results in a permanent genetic change for the whole organism and subsequent generations. This "germ line gene therapy" is considered by many to be unethical in human beings. The second type of gene therapy, "somatic cell gene therapy", is analogous to an organ transplant. In this case, one or more specific tissues are targeted by direct treatment or by removal of the tissue, addition of the therapeutic gene or genes in the laboratory, and return of the treated cells to the patient. Clinical trials of somatic cell gene therapy began in the late 1990s, mostly for the treatment of cancers and blood, liver, and lung disorders.

Despite a great deal of publicity and promises, the history of human gene therapy has been characterized by relatively limited success. The effect of introducing a gene into cells often promotes only partial and/or transient relief from the symptoms of the disease being treated. Some gene therapy trial patients have suffered adverse consequences of the treatment itself, including deaths. In some cases, the adverse effects result from disruption of essential genes within the patient's genome by insertional inactivation. In others, viral vectors used for gene therapy have been contaminated with infectious virus. Nevertheless, gene therapy is still held to be a promising future area of medicine, and is an area where there is a significant level of research and development activity.

Component of Molecular Cloning

Recombinant DNA

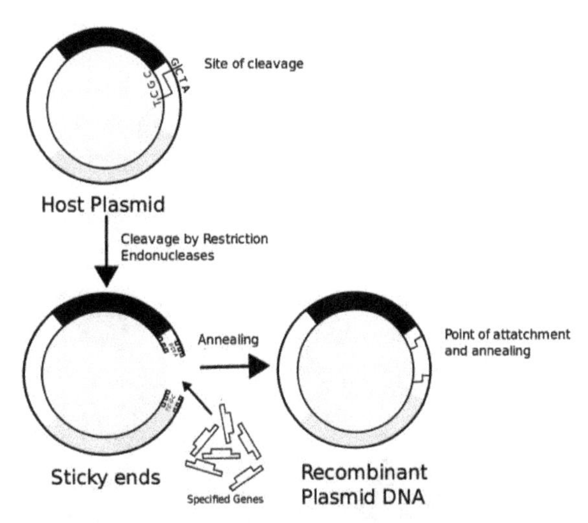

Construction of recombinant DNA, in which a foreign DNA fragment is inserted into a plasmid vector. In this example, the gene indicated by the white color is inactivated upon insertion of the foreign DNA fragment.

Recombinant DNA (rDNA) molecules are DNA molecules formed by laboratory methods of genetic recombination (such as molecular cloning) to bring together genetic material from multiple sources, creating sequences that would not otherwise be found in the genome. Recombinant DNA is possible because DNA molecules from all organisms share the same chemical structure. They differ only in the nucleotide sequence within that identical overall structure.

Introduction

Recombinant DNA is the general name for a piece of DNA that has been created by the combination of at least two strands. Recombinant DNA molecules are sometimes called **chimeric DNA**, because they can be made of material from two different species, like the mythical chimera. R-DNA technology uses palindromic sequences and leads to the production of sticky and blunt ends.

The DNA sequences used in the construction of recombinant DNA molecules can originate from any species. For example, plant DNA may be joined to bacterial DNA, or human DNA may be joined with fungal DNA. In addition, DNA sequences that do not occur anywhere in nature may be created by the chemical synthesis of DNA, and incorporated into recombinant molecules. Using recombinant DNA technology and synthetic DNA, literally any DNA sequence may be created and introduced into any of a very wide range of living organisms.

Proteins that can result from the expression of recombinant DNA within living cells are termed recombinant proteins. When recombinant DNA encoding a protein is introduced into a host organism, the recombinant protein is not necessarily produced. Expression of foreign proteins requires the use of specialized expression vectors and often necessitates significant restructuring by foreign coding sequences.

Recombinant DNA differs from genetic recombination in that the former results from artificial methods in the test tube, while the latter is a normal biological process that results in the remixing of existing DNA sequences in essentially all organisms.

Creating Recombinant DNA

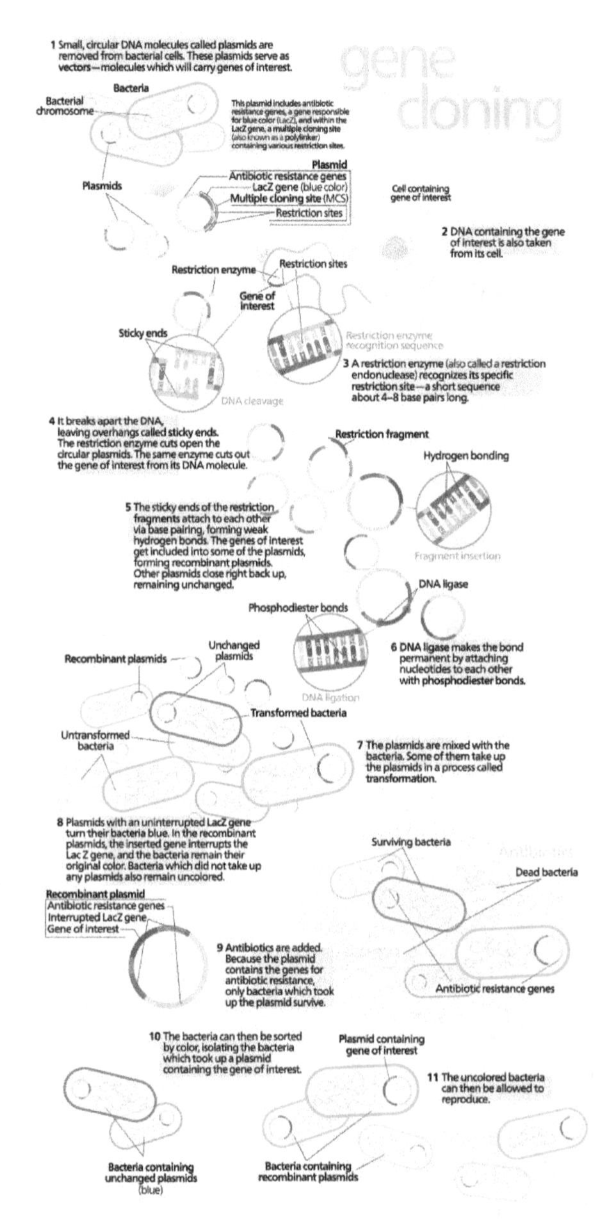

Molecular cloning is the laboratory process used to create recombinant DNA. It is one of two widely used methods, along with polymerase chain reaction (PCR) used to direct the replication of any specific DNA sequence chosen by the experimentalist. The fundamental difference between the two methods is that molecular cloning involves replication of the DNA within a living cell, while PCR replicates DNA in the test tube, free of living cells.

Formation of recombinant DNA requires a cloning vector, a DNA molecule that replicates within a living cell. Vectors are generally derived from plasmids or viruses, and represent relatively small segments of DNA that contain necessary genetic signals for replication, as well as additional elements for convenience in inserting foreign DNA, identifying cells that contain recombinant DNA, and, where appropriate, expressing the foreign DNA. The choice of vector for molecular cloning depends on the choice of host organism, the size of the DNA to be cloned, and whether and how the foreign DNA is to be expressed. The DNA segments can be combined by using a variety of methods, such as restriction enzyme/ligase cloning or Gibson assembly.

In standard cloning protocols, the cloning of any DNA fragment essentially involves seven steps: (1) Choice of host organism and cloning vector, (2) Preparation of vector DNA, (3) Preparation of DNA to be cloned, (4) Creation of recombinant DNA, (5) Introduction of recombinant DNA into the host organism, (6) Selection of organisms containing recombinant DNA, and (7) Screening for clones with desired DNA inserts and biological properties. *These steps are described in some detail in a related article (molecular cloning).*

Expression of Recombinant DNA

Following transplantation into the host organism, the foreign DNA contained within the recombinant DNA construct may or may not be expressed. That is, the DNA may simply be replicated without expression, or it may be transcribed and translated at a recombinant protein is produced. Generally speaking, expression of a foreign gene requires restructuring the gene to include sequences that are required for producing an mRNA molecule that can be used by the host's translational apparatus (e.g. promoter, translational initiation signal, and transcriptional terminator). Specific changes to the host organism may be made to improve expression of the ectopic gene. In addition, changes may be needed to the coding sequences as well, to optimize translation, make the protein soluble, direct the recombinant protein to the proper cellular or extracellular location, and stabilize the protein from degradation.

Properties of Organisms Containing Recombinant DNA

In most cases, organisms containing recombinant DNA have apparently normal phenotypes. That is, their appearance, behavior and metabolism are usually unchanged, and the only way to demonstrate the presence of recombinant sequences is to examine the DNA itself, typically using a polymerase chain reaction (PCR) test. Significant exceptions exist, and are discussed below.

If the rDNA sequences encode a gene that is expressed, then the presence of RNA and/or protein products of the recombinant gene can be detected, typically using RT-PCR or western hybridization methods. Gross phenotypic changes are not the norm, unless the recombinant gene has been chosen and modified so as to generate biological activity in the host organism. Additional phenotypes that are encountered include toxicity to the host organism induced by the recombinant gene product, especially if it is over-expressed or expressed within inappropriate cells or tissues.

In some cases, recombinant DNA can have deleterious effects even if it is not expressed. One mechanism by which this happens is insertional inactivation, in which the rDNA becomes inserted into a host cell's gene. In some cases, researchers use this phenomenon to "knock out" genes to determine their biological function and importance. Another mechanism by which rDNA insertion into chromosomal DNA can affect gene expression is by inappropriate activation of previously unexpressed host cell genes. This can happen, for example, when a recombinant DNA fragment containing an active promoter becomes located next to a previously silent host cell gene, or when a host cell gene that functions to restrain gene expression undergoes insertional inactivation by recombinant DNA.

Applications of Recombinant DNA Technology

Recombinant DNA is widely used in biotechnology, medicine and research. Today, recombinant proteins and other products that result from the use of rDNA technology are found in essentially every western pharmacy, doctor's or veterinarian's office, medical testing laboratory, and biological research laboratory. In addition, organisms that have been manipulated using recombinant DNA technology, as well as products derived from those organisms, have found their way into many farms, supermarkets, home medicine cabinets, and even pet shops, such as those that sell GloFish and other genetically modified animals.

A group of GloFish fluorescent fish

The most common application of recombinant DNA is in basic research, in which the technology is important to most current work in the biological and biomedical sciences. Recombinant DNA is used to identify, map and sequence genes, and to determine their function. rDNA probes are employed in analyzing gene expression within individual cells, and throughout the tissues of whole organisms. Recombinant proteins are widely used as reagents in laboratory experiments and to generate antibody probes for examining protein synthesis within cells and organisms.

Many additional practical applications of recombinant DNA are found in industry, food production, human and veterinary medicine, agriculture, and bioengineering. Some specific examples are identified below.

Recombinant chymosin

Found in rennet, chymosin is an enzyme required to manufacture cheese. It was the first genetically engineered food additive used commercially. Traditionally, processors obtained chymosin from rennet, a preparation derived from the fourth stomach of milk-fed calves. Scientists engineered a non-pathogenic strain (K-12) of *E. coli* bacteria for large-scale laboratory production of the enzyme. This microbiologically produced recombinant enzyme, identical structurally to the calf derived enzyme, costs less and is produced in abundant quantities. Today about 60% of U.S. hard cheese is made with genetically engineered chymosin. In 1990, FDA granted chymosin "generally recognized as safe" (GRAS) status based on data showing that the enzyme was safe.

Recombinant human insulin

Almost completely replaced insulin obtained from animal sources (e.g. pigs and cattle) for the treatment of insulin-dependent diabetes. A variety of different recombinant insulin preparations are in widespread use. Recombinant insulin is synthesized by inserting the human insulin gene into *E. coli*, or yeast (saccharomyces cerevisiae) which then produces insulin for human use.

Recombinant human growth hormone (HGH, somatotropin)

Administered to patients whose pituitary glands generate insufficient quantities to support normal growth and development. Before recombinant HGH became available, HGH for therapeutic use was obtained from pituitary glands of cadavers. This unsafe practice led to some patients developing Creutzfeldt–Jakob disease. Recombinant HGH eliminated this problem, and is now used therapeutically. It has also been misused as a performance-enhancing drug by athletes and others. DrugBank entry

Recombinant blood clotting factor VIII

A blood-clotting protein that is administered to patients with forms of the bleeding disorder hemophilia, who are unable to produce factor VIII in quantities sufficient to support normal blood coagulation. Before the development of recombinant factor VIII, the protein was obtained by processing large quantities of human blood from multiple donors, which carried a very high risk of transmission of blood borne infectious diseases, for example HIV and hepatitis B. DrugBank entry

Recombinant hepatitis B vaccine

Hepatitis B infection is controlled through the use of a recombinant hepatitis B vaccine, which contains a form of the hepatitis B virus surface antigen that is produced in yeast cells. The development of the recombinant subunit vaccine was an important and necessary development because hepatitis B virus, unlike other common viruses such as polio virus, cannot be grown in vitro. Vaccine information from Hepatitis B Foundation

Diagnosis of infection with HIV

Each of the three widely used methods for diagnosing HIV infection has been developed using recombinant DNA. The antibody test (ELISA or western blot) uses a recombinant HIV protein to test for the presence of antibodies that the body has produced in response to an HIV infection. The DNA test looks for the presence of HIV genetic material using reverse transcription polymerase chain reaction (RT-PCR). Development of the RT-PCR test was made possible by the molecular cloning and sequence analysis of HIV genomes. HIV testing page from US Centers for Disease Control (CDC)

Golden rice

A recombinant variety of rice that has been engineered to express the enzymes responsible for β-carotene biosynthesis. This variety of rice holds substantial promise for reducing the incidence of vitamin A deficiency in the world's population. Golden rice is not currently in use, pending the resolution of regulatory and intellectual property issues.

Herbicide-resistant crops

Commercial varieties of important agricultural crops (including soy, maize/corn, sorghum, canola, alfalfa and cotton) have been developed that incorporate a recombinant gene that results in resistance to the herbicide glyphosate (trade name *Roundup*), and simplifies weed control by glyphosate application. These crops are in common commercial use in several countries.

Insect-resistant crops

Bacillus thuringeiensis is a bacterium that naturally produces a protein (Bt toxin) with insecticidal properties. The bacterium has been applied to crops as an insect-control strategy for many years, and this practice has been widely adopted in agriculture and gardening. Recently, plants have been developed that express a recombinant form of the bacterial protein, which may effectively control some insect predators. Environmental issues associated with the use of these transgenic crops have not been fully resolved.

History of Recombinant DNA

The idea of recombinant DNA was first proposed by Peter Lobban, a graduate student of Prof. Dale Kaiser in the Biochemistry Department at Stanford University Medical School. The first publications describing the successful production and intracellular replication of recombinant DNA appeared in 1972 and 1973. Stanford University applied for a US patent on recombinant DNA in 1974, listing the inventors as Stanley N. Cohen and Herbert W. Boyer; this patent was awarded in 1980. The first licensed drug generated using recombinant DNA technology was human insulin, developed by Genentech and Licensed by Eli Lilly and Company.

Controversy

Scientists associated with the initial development of recombinant DNA methods recognized that the potential existed for organisms containing recombinant DNA to have undesirable or dangerous properties. At the 1975 Asilomar Conference on Recombinant DNA, these concerns were discussed and a voluntary moratorium on recombinant DNA research was initiated for experiments that were considered particularly risky. This moratorium was widely observed until the National Institutes of Health (USA) developed and issued formal guidelines for rDNA work. Today, recombinant DNA molecules and recombinant proteins are usually not regarded as dangerous. However, concerns remain about some organisms that express recombinant DNA, particularly when they leave the laboratory and are introduced into the environment or food chain. These concerns are discussed in the articles on genetically modified organisms and genetically modified food controversies.

Polymerase Chain Reaction

The polymerase chain reaction (PCR) is a technique used in molecular biology to amplify a single copy or a few copies of a piece of DNA across several orders of magnitude, generating thousands to millions of copies of a particular DNA sequence. It is an easy and cheap tool to amplify a focused segment of DNA, useful in the diagnosis and monitoring of genetic diseases, identification of criminals (under the field of forensics), studying the function of targeted segment, etc.

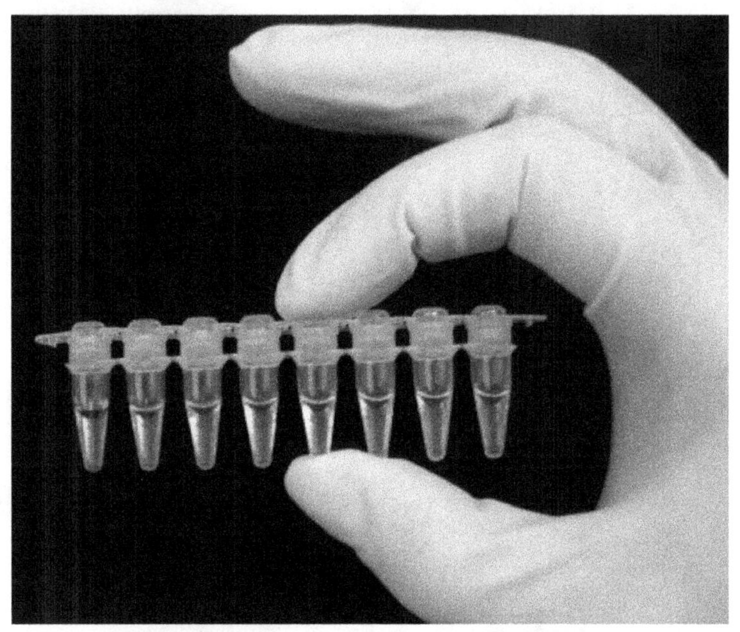

A strip of eight PCR tubes, each containing a 100 μl reaction mixture

Developed in 1983 by Kary Mullis, PCR is now a common and often indispensable technique used in medical and biological research labs for a variety of applications. These include DNA cloning for sequencing, DNA-based phylogeny, or functional analysis of genes; the diagnosis of hereditary diseases; the identification of genetic fingerprints (used in forensic sciences and DNA paternity testing); and the detection and diagnosis of infectious diseases. In 1993, Mullis was awarded the Nobel Prize in Chemistry along with Michael Smith for his work on PCR.

The method relies on thermal cycling, consisting of cycles of repeated heating and cooling of the reaction for DNA melting and enzymatic replication of the DNA. Primers (short DNA fragments) containing sequences complementary to the target region along with a DNA polymerase, which the method is named after, are key components to enable selective and repeated amplification. As PCR progresses, the DNA generated is itself used as a template for replication, setting in motion a chain reaction in which the DNA template is exponentially amplified. PCR can be extensively modified to perform a wide array of genetic manipulations.

Almost all PCR applications employ a heat-stable DNA polymerase, such as Taq polymerase (an enzyme originally isolated from the bacterium *Thermus aquaticus*). This DNA polymerase enzymatically assembles a new DNA strand from DNA building-blocks, the nucleotides, by using single-stranded DNA as a template and DNA oligonucleotides (also called DNA primers), which are required for initiation of DNA synthesis. The vast majority of PCR methods use thermal cycling, i.e., alternately heating and cooling the PCR sample through a defined series of temperature steps.

In the first step, the two strands of the DNA double helix are physically separated at a high temperature in a process called DNA melting. In the second step, the temperature is lowered and the two DNA strands become templates for DNA polymerase to selectively amplify the target DNA.

The selectivity of PCR results from the use of primers that are complementary to the DNA region targeted for amplification under specific thermal cycling conditions.

Placing a strip of eight PCR tubes into a thermal cycler

Principles

A thermal cycler for PCR

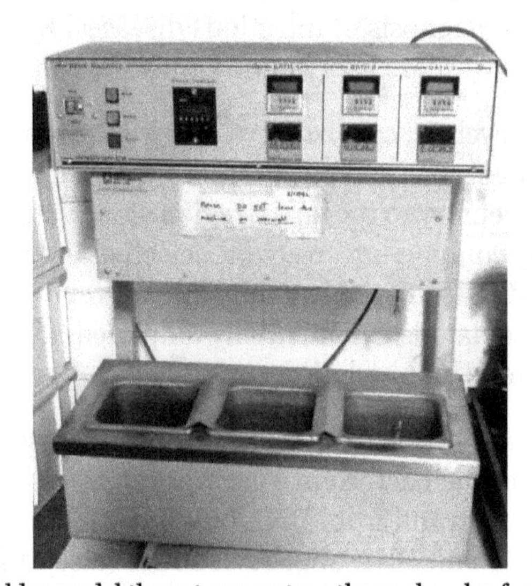

An older model three-temperature thermal cycler for PCR

PCR amplifies a specific region of a DNA strand (the DNA target). Most PCR methods typically amplify DNA fragments of between 0.1 and 10 kilo base pairs (kbp), although some techniques allow for amplification of fragments up to 40 kbp in size. The amount of amplified product is determined by the available substrates in the reaction, which become limiting as the reaction progresses.

A basic PCR set up requires several components and reagents. These components include:

- *DNA template* that contains the DNA region (target) to amplify
- Two *primers* that are complementary to the 3' (three prime) ends of each of the sense and anti-sense strand of the DNA target. Primers can be custom made in a laboratory that are complementary to the DNA segment to be amplified.
- *Taq polymerase* or another DNA polymerase with a temperature optimum at around 70 °C. DNA Polymerase should be able to elongate the DNA strand with the help of Primers, as the DNA Polymerase can not attach to a DNA strand and elongate on its own. It should also be heat resistant, so that it can withstand the denaturation process.
- *Deoxynucleoside triphosphates* (dNTPs, sometimes called "deoxynucleotide triphosphates"; nucleotides containing triphosphate groups), the building-blocks from which the DNA polymerase synthesizes a new DNA strand.
- *Buffer solution*, providing a suitable chemical environment for optimum activity and stability of the DNA polymerase
- *Bivalent cations*, magnesium or manganese ions; generally Mg^{2+} is used, but Mn^{2+} can be used for PCR-mediated DNA mutagenesis, as higher Mn^{2+} concentration increases the error rate during DNA synthesis
- *Monovalent cation* potassium ions

The PCR is commonly carried out in a reaction volume of 10–200 μl in small reaction tubes (0.2–0.5 ml volumes) in a thermal cycler. The thermal cycler heats and cools the reaction tubes to achieve the temperatures required at each step of the reaction. Many modern thermal cyclers make use of the Peltier effect, which permits both heating and cooling of the block holding the PCR tubes simply by reversing the electric current. Thin-walled reaction tubes permit favorable thermal conductivity to allow for rapid thermal equilibration. Most thermal cyclers have heated lids to prevent condensation at the top of the reaction tube. Older thermocyclers lacking a heated lid require a layer of oil on top of the reaction mixture or a ball of wax inside the tube.

Procedure

Typically, PCR consists of a series of 20–40 repeated temperature changes, called cycles, with each cycle commonly consisting of 2–3 discrete temperature steps, usually three (Figure below). The cycling is often preceded by a single temperature step at a high temperature (>90 °C), and followed by one hold at the end for final product extension or brief storage. The temperatures used and the length of time they are applied in each cycle depend on a variety of parameters. These include the enzyme used for DNA synthesis, the concentration of divalent ions and dNTPs in the reaction, and the melting temperature (Tm) of the primers.

- *Initialization step* (Only required for DNA polymerases that require heat activation by hot-start PCR.): This step consists of heating the reaction to a temperature of 94–96 °C (or 98 °C if extremely thermostable polymerases are used), which is held for 1–9 minutes.

- *Denaturation step*: This step is the first regular cycling event and consists of heating the reaction to 94–98 °C for 20–30 seconds. It causes DNA melting of the DNA template by disrupting the hydrogen bonds between complementary bases, yielding single-stranded DNA molecules.

- *Annealing step*: The reaction temperature is lowered to 50–65 °C for 20–40 seconds allowing annealing of the primers to the single-stranded DNA template. This temperature must be low enough to allow for hybridization of the primer to the strand, but high enough for the hybridization to be specific, i.e., the primer should only bind to a perfectly complementary part of the template. If the temperature is too low, the primer could bind imperfectly. If it is too high, the primer might not bind. Typically the annealing temperature is about 3–5 °C below the Tm of the primers used. Stable DNA–DNA hydrogen bonds are only formed when the primer sequence very closely matches the template sequence. The polymerase binds to the primer-template hybrid and begins DNA formation. It is very vital to determine the annealing temperature in PCR. This is because in PCR, efficiency and specificity are affected by the annealing temperature. An incorrect annealing temperature will cause an error in the test.

- *Extension/elongation step*: The temperature at this step depends on the DNA polymerase used; Taq polymerase has its optimum activity temperature at 75–80 °C, and commonly a temperature of 72 °C is used with this enzyme. At this step the DNA polymerase synthesizes a new DNA strand complementary to the DNA template strand by adding dNTPs that are complementary to the template in 5' to 3' direction, condensing the 5'-phosphate group of the dNTPs with the 3'-hydroxyl group at the end of the nascent (extending) DNA strand. The extension time depends both on the DNA polymerase used and on the length of the DNA fragment to amplify. As a rule-of-thumb, at its optimum temperature, the DNA polymerase polymerizes a thousand bases per minute. Under optimum conditions, i.e., if there are no limitations due to limiting substrates or reagents, at each extension step, the amount of DNA target is doubled, leading to exponential (geometric) amplification of the specific DNA fragment.

The processes of denaturation, annealing and elongation constitute of one cycle. Multiple cycles are required to amplify DNA. The formula used to calculate the number of DNA copies formed after a given # of cycles repeated is, 2^# of cycles.

- *Final elongation*: This single step is occasionally performed at a temperature of 70–74 °C (this is the temperature needed for optimal activity for most polymerases used in PCR) for 5–15 minutes after the last PCR cycle to ensure that any remaining single-stranded DNA is fully extended.

- *Final hold*: This step at 4–15 °C for an indefinite time may be employed for short-term storage of the reaction.

To check whether the PCR generated the anticipated DNA fragment (also sometimes referred to as the amplimer or amplicon), agarose gel electrophoresis is employed for size separation of the PCR products. The size(s) of PCR products is determined by comparison with a DNA ladder (a molecular weight marker), which contains DNA fragments of known size, run on the gel alongside the PCR products.

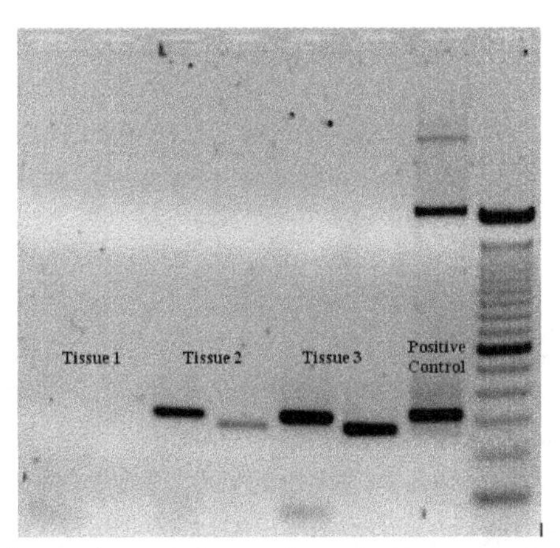

Ethidium bromide-stained PCR products after gel electrophoresis. Two sets of primers were used to amplify a target sequence from three different tissue samples. No amplification is present in sample #1; DNA bands in sample #2 and #3 indicate successful amplification of the target sequence. The gel also shows a positive control, and a DNA ladder containing DNA fragments of defined length for sizing the bands in the experimental PCRs.

Stages

The PCR process can be divided into three stages:

Exponential amplification: At every cycle, the amount of product is doubled (assuming 100% reaction efficiency). The reaction is very sensitive: only minute quantities of DNA must be present.

Leveling off stage: The reaction slows as the DNA polymerase loses activity and as consumption of reagents such as dNTPs and primers causes them to become limiting.

Plateau: No more product accumulates due to exhaustion of reagents and enzyme.

Optimization

In practice, PCR can fail for various reasons, in part due to its sensitivity to contamination causing amplification of spurious DNA products. Because of this, a number of techniques and procedures have been developed for optimizing PCR conditions. Contamination with extraneous DNA is addressed with lab protocols and procedures that separate pre-PCR mixtures from potential DNA contaminants. This usually involves spatial separation of PCR-setup areas from areas for analysis or purification of PCR products, use of disposable plasticware, and thoroughly cleaning the work surface between reaction setups. Primer-design techniques are important in improving PCR product yield and in avoiding the formation of spurious products, and the usage of alternate buffer components or polymerase enzymes can help with amplification of long or otherwise problematic regions of DNA. Addition of reagents, such as formamide, in buffer systems may increase the specificity and yield of PCR. Computer simulations of theoretical PCR results (Electronic PCR) may be performed to assist in primer design.

Applications

Selective DNA Isolation

PCR allows isolation of DNA fragments from genomic DNA by selective amplification of a specific region of DNA. This use of PCR augments many methods, such as generating hybridization probes for Southern or northern hybridization and DNA cloning, which require larger amounts of DNA, representing a specific DNA region. PCR supplies these techniques with high amounts of pure DNA, enabling analysis of DNA samples even from very small amounts of starting material.

Other applications of PCR include DNA sequencing to determine unknown PCR-amplified sequences in which one of the amplification primers may be used in Sanger sequencing, isolation of a DNA sequence to expedite recombinant DNA technologies involving the insertion of a DNA sequence into a plasmid, phage, or cosmid (depending on size) or the genetic material of another organism. Bacterial colonies *(such as E. coli)* can be rapidly screened by PCR for correct DNA vector constructs. PCR may also be used for genetic fingerprinting; a forensic technique used to identify a person or organism by comparing experimental DNAs through different PCR-based methods.

Some PCR 'fingerprints' methods have high discriminative power and can be used to identify genetic relationships between individuals, such as parent-child or between siblings, and are used in paternity testing (Fig. 4). This technique may also be used to determine evolutionary relationships among organisms when certain molecular clocks are used (i.e., the 16S rRNA and recA genes of microorganisms).

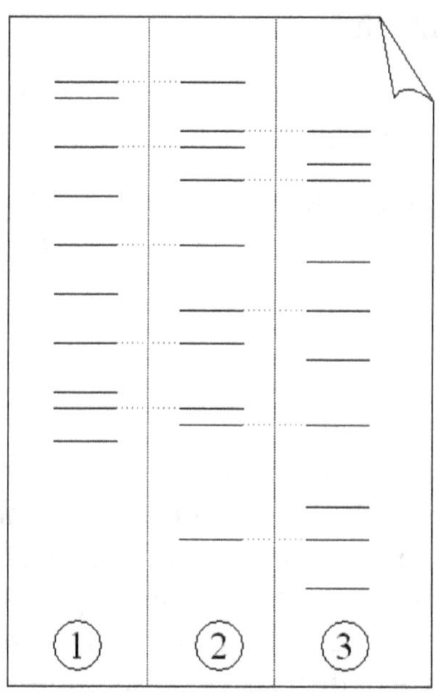

Electrophoresis of PCR-amplified DNA fragments. (1) Father. (2) Child. (3) Mother. The child has inherited some, but not all of the fingerprint of each of its parents, giving it a new, unique fingerprint.

Amplification and Quantification of DNA

Because PCR amplifies the regions of DNA that it targets, PCR can be used to analyze extremely small amounts of sample. This is often critical for forensic analysis, when only a trace amount of DNA is available as evidence. PCR may also be used in the analysis of ancient DNA that is tens of thousands of years old. These PCR-based techniques have been successfully used on animals, such as a forty-thousand-year-old mammoth, and also on human DNA, in applications ranging from the analysis of Egyptian mummies to the identification of a Russian tsar and the body of English king Richard III.

Quantitative PCR methods allow the estimation of the amount of a given sequence present in a sample—a technique often applied to quantitatively determine levels of gene expression. Quantitative PCR is an established tool for DNA quantification that measures the accumulation of DNA product after each round of PCR amplification.

Disease Diagnosis

PCR permits early diagnosis of malignant diseases such as leukemia and lymphomas, which is currently the highest-developed in cancer research and is already being used routinely. PCR assays can be performed directly on genomic DNA samples to detect translocation-specific malignant cells at a sensitivity that is at least 10,000 fold higher than that of other methods.

PCR allows for rapid and highly specific diagnosis of infectious diseases, including those caused by bacteria or viruses. PCR also permits identification of non-cultivatable or slow-growing microorganisms such as mycobacteria, anaerobic bacteria, or viruses from tissue culture assays and animal models. The basis for PCR diagnostic applications in microbiology is the detection of infectious agents and the discrimination of non-pathogenic from pathogenic strains by virtue of specific genes.

Viral DNA can likewise be detected by PCR. The primers used must be specific to the targeted sequences in the DNA of a virus, and PCR can be used for diagnostic analyses or DNA sequencing of the viral genome. The high sensitivity of PCR permits virus detection soon after infection and even before the onset of disease. Such early detection may give physicians a significant lead time in treatment. The amount of virus ("viral load") in a patient can also be quantified by PCR-based DNA quantitation techniques.

Limitations

DNA polymerase is prone to error, which in turn causes mutations in the PCR fragments that are made. Additionally, the specificity of the PCR fragments can mutate to the template DNA, due to nonspecific binding of primers. Furthermore, prior information on the sequence is necessary in order to generate the primers.

Variations

- *Allele-speci ic PCR*: a diagnostic or cloning technique based on single-nucleotide variations (single-base differences in a patient). It requires prior knowledge of a DNA sequence, including differences between alleles, and uses primers whose 3' ends encompass the SNV (base pair buffer around SNV usually incorporated). PCR amplification under stringent conditions is much less efficient in the presence of a mismatch between template and primer, so successful amplification with an SNP-specific primer signals presence of the specific SNP in a sequence. SNP genotyping for more information.

- *Assembly PCR* or *Polymerase Cycling Assembly (PCA)*: artificial synthesis of long DNA sequences by performing PCR on a pool of long oligonucleotides with short overlapping segments. The oligonucleotides alternate between sense and antisense directions, and the overlapping segments determine the order of the PCR fragments, thereby selectively producing the final long DNA product.

- *Asymmetric PCR*: preferentially amplifies one DNA strand in a double-stranded DNA template. It is used in sequencing and hybridization probing where amplification of only one of the two complementary strands is required. PCR is carried out as usual, but with a great excess of the primer for the strand targeted for amplification. Because of the slow (arithmetic) amplification later in the reaction after the limiting primer has been used up, extra cycles of PCR are required. A recent modification on this process, known as *L*inear-*A*fter-*T*he-*E*xponential-PCR (LATE-PCR), uses a limiting primer with a higher melting temperature (Tm) than the excess primer to maintain reaction efficiency as the limiting primer concentration decreases mid-reaction.

- *Dial-out PCR*: a highly parallel method for retrieving accurate DNA molecules for gene synthesis. A complex library of DNA molecules is modified with unique flanking tags before massively parallel sequencing. Tag-directed primers then enable the retrieval of molecules with desired sequences by PCR.

- *Digital PCR (dPCR)*: used to measure the quantity of a target DNA sequence in a DNA sample. The DNA sample is highly diluted so that after running many PCRs in parallel, some of them do not receive a single molecule of the target DNA. The target DNA concentration is calculated using the proportion of negative outcomes. Hence the name 'digital PCR'.

- *Helicase-dependent amplification*: similar to traditional PCR, but uses a constant temperature rather than cycling through denaturation and annealing/extension cycles. DNA helicase, an enzyme that unwinds DNA, is used in place of thermal denaturation.

- *Hot start PCR*: a technique that reduces non-specific amplification during the initial set up stages of the PCR. It may be performed manually by heating the reaction components to the denaturation temperature (e.g., 95 °C) before adding the polymerase. Specialized enzyme systems have been developed that inhibit the polymerase's activity at ambient temperature, either by the binding of an antibody or by the presence of covalently bound inhibitors that dissociate only after a high-temperature activation step. Hot-start/cold-finish PCR is achieved with new hybrid polymerases that are inactive at ambient temperature and are instantly activated at elongation temperature.

- *In silico PCR* (digital PCR, virtual PCR, electronic PCR, e-PCR) refers to computational tools used to calculate theoretical polymerase chain reaction results using a given set of primers (probes) to amplify DNA sequences from a sequenced genome or transcriptome. In silico PCR was proposed as an educational tool for molecular biology.

- *Intersequence-specific PCR* (ISSR): a PCR method for DNA fingerprinting that amplifies regions between simple sequence repeats to produce a unique fingerprint of amplified fragment lengths.

- *Inverse PCR*: is commonly used to identify the flanking sequences around genomic inserts. It involves a series of DNA digestions and self ligation, resulting in known sequences at either end of the unknown sequence.

- *Ligation-mediated PCR*: uses small DNA linkers ligated to the DNA of interest and multiple primers annealing to the DNA linkers; it has been used for DNA sequencing, genome walking, and DNA footprinting.
- *Methylation-specific PCR* (MSP): developed by Stephen Baylin and Jim Herman at the Johns Hopkins School of Medicine, and is used to detect methylation of CpG islands in genomic DNA. DNA is first treated with sodium bisulfite, which converts unmethylated cytosine bases to uracil, which is recognized by PCR primers as thymine. Two PCRs are then carried out on the modified DNA, using primer sets identical except at any CpG islands within the primer sequences. At these points, one primer set recognizes DNA with cytosines to amplify methylated DNA, and one set recognizes DNA with uracil or thymine to amplify unmethylated DNA. MSP using qPCR can also be performed to obtain quantitative rather than qualitative information about methylation.
- *Miniprimer PCR*: uses a thermostable polymerase (S-Tbr) that can extend from short primers ("smalligos") as short as 9 or 10 nucleotides. This method permits PCR targeting to smaller primer binding regions, and is used to amplify conserved DNA sequences, such as the 16S (or eukaryotic 18S) rRNA gene.
- *Multiplex ligation-dependent probe amplification* (*MLPA*): permits amplifying multiple targets with a single primer pair, thus avoiding the resolution limitations of multiplex PCR.
- *Multiplex-PCR*: consists of multiple primer sets within a single PCR mixture to produce amplicons of varying sizes that are specific to different DNA sequences. By targeting multiple genes at once, additional information may be gained from a single test-run that otherwise would require several times the reagents and more time to perform. Annealing temperatures for each of the primer sets must be optimized to work correctly within a single reaction, and amplicon sizes. That is, their base pair length should be different enough to form distinct bands when visualized by gel electrophoresis.
- *Nanoparticle-Assisted PCR (nanoPCR)*: In recent years, it has been reported that some nanoparticles (NPs) can enhance the efficiency of PCR (thus being called nanoPCR), and some even perform better than the original PCR enhancers. It was also found that quantum dots (QDs) can improve PCR specificity and efficiency. Single-walled carbon nanotubes (SWCNTs) and multi-walled carbon nanotubes (MWCNTs) are efficient in enhancing the amplification of long PCR. Carbon nanopowder (CNP) was reported be able to improve the efficiency of repeated PCR and long PCR. ZnO, TiO2, and Ag NPs were also found to increase PCR yield. Importantly, already known data has indicated that non-metallic NPs retained acceptable amplification fidelity. Given that many NPs are capable of enhancing PCR efficiency, it is clear that there is likely to be great potential for nanoPCR technology improvements and product development.
- *Nested PCR*: increases the specificity of DNA amplification, by reducing background due to non-specific amplification of DNA. Two sets of primers are used in two successive PCRs. In the first reaction, one pair of primers is used to generate DNA products, which besides the intended target, may still consist of non-specifically amplified DNA fragments. The pro-

duct(s) are then used in a second PCR with a set of primers whose binding sites are completely or partially different from and located 3' of each of the primers used in the first reaction. Nested PCR is often more successful in specifically amplifying long DNA fragments than conventional PCR, but it requires more detailed knowledge of the target sequences.

- *Overlap-extension PCR* or *Splicing by overlap extension (SOEing)* : a genetic engineering technique that is used to splice together two or more DNA fragments that contain complementary sequences. It is used to join DNA pieces containing genes, regulatory sequences, or mutations; the technique enables creation of specific and long DNA constructs. It can also introduce deletions, insertions or point mutations into a DNA sequence.
- *PAN-AC*: uses isothermal conditions for amplification, and may be used in living cells.
- *quantitative PCR* (qPCR): used to measure the quantity of a target sequence (commonly in real-time). It quantitatively measures starting amounts of DNA, cDNA, or RNA. quantitative PCR is commonly used to determine whether a DNA sequence is present in a sample and the number of its copies in the sample. *Quantitative PCR* has a very high degree of precision. Quantitative PCR methods use fluorescent dyes, such as Sybr Green, EvaGreen or fluorophore-containing DNA probes, such as TaqMan, to measure the amount of amplified product in real time. It is also sometimes abbreviated to RT-PCR (*real-time* PCR) but this abbreviation should be used only for reverse transcription PCR. qPCR is the appropriate contractions for quantitative PCR (real-time PCR).
- *Reverse Transcription PCR (RT-PCR)*: for amplifying DNA from RNA. Reverse transcriptase reverse transcribes RNA into cDNA, which is then amplified by PCR. RT-PCR is widely used in expression profiling, to determine the expression of a gene or to identify the sequence of an RNA transcript, including transcription start and termination sites. If the genomic DNA sequence of a gene is known, RT-PCR can be used to map the location of exons and introns in the gene. The 5' end of a gene (corresponding to the transcription start site) is typically identified by RACE-PCR (*Rapid Amplification of cDNA Ends*).
- *Solid Phase PCR*: encompasses multiple meanings, including Polony Amplification (where PCR colonies are derived in a gel matrix, for example), Bridge PCR (primers are covalently linked to a solid-support surface), conventional Solid Phase PCR (where Asymmetric PCR is applied in the presence of solid support bearing primer with sequence matching one of the aqueous primers) and Enhanced Solid Phase PCR (where conventional Solid Phase PCR can be improved by employing high Tm and nested solid support primer with optional application of a thermal 'step' to favour solid support priming).
- *Suicide PCR*: typically used in paleogenetics or other studies where avoiding false positives and ensuring the specificity of the amplified fragment is the highest priority. It was originally described in a study to verify the presence of the microbe Yersinia pestis in dental samples obtained from 14th Century graves of people supposedly killed by plague during the medieval Black Death epidemic. The method prescribes the use of any primer combination only once in a PCR (hence the term "suicide"), which should never have been used in any positive control PCR reaction, and the primers should always target a genomic region never amplified before in the lab using this or any other set of primers. This ensures that

no contaminating DNA from previous PCR reactions is present in the lab, which could otherwise generate false positives.

- *Thermal asymmetric interlaced PCR (TAIL-PCR)*: for isolation of an unknown sequence flanking a known sequence. Within the known sequence, TAIL-PCR uses a nested pair of primers with differing annealing temperatures; a degenerate primer is used to amplify in the other direction from the unknown sequence.

- *Touchdown PCR* (*Step-down PCR*): a variant of PCR that aims to reduce nonspecific background by gradually lowering the annealing temperature as PCR cycling progresses. The annealing temperature at the initial cycles is usually a few degrees (3–5 °C) above the T_m of the primers used, while at the later cycles, it is a few degrees (3–5 °C) below the primer T_m. The higher temperatures give greater specificity for primer binding, and the lower temperatures permit more efficient amplification from the specific products formed during the initial cycles.

- *Universal Fast Walking*: for genome walking and genetic fingerprinting using a more specific 'two-sided' PCR than conventional 'one-sided' approaches (using only one gene-specific primer and one general primer—which can lead to artefactual 'noise') by virtue of a mechanism involving lariat structure formation. Streamlined derivatives of UFW are LaNe RAGE (lariat-dependent nested PCR for rapid amplification of genomic DNA ends), 5'RACE LaNe and 3'RACE LaNe.

History

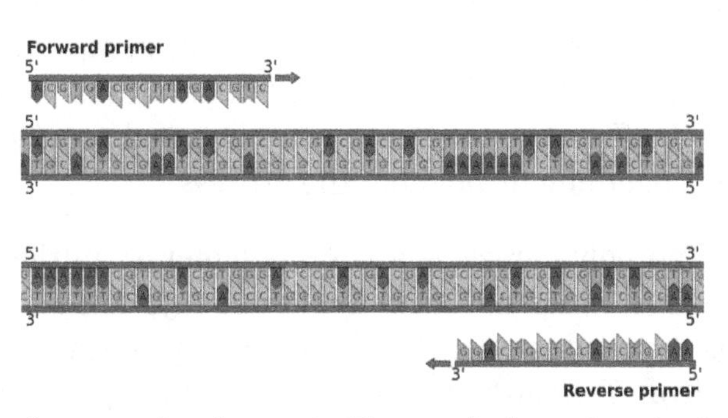

Diagrammatic representation of an example primer pair. The use of primers in an in vitro assay to allow DNA synthesis was a major innovation that allowed the development of PCR.

A 1971 paper in the Journal of Molecular Biology by Kjell Kleppe and co-workers in the laboratory of H. Gobind Khorana first described a method using an enzymatic assay to replicate a short DNA template with primers *in vitro*. However, this early manifestation of the basic PCR principle did not receive much attention at the time, and the invention of the polymerase chain reaction in 1983 is generally credited to Kary Mullis.

When Mullis developed the PCR in 1983, he was working in Emeryville, California for Cetus Corporation, one of the first biotechnology companies. There, he was responsible for synthesizing short chains of DNA. Mullis has written that he conceived of PCR while cruising along the Pacific Coast Highway one night in his car. He was playing in his mind with a new way of analyzing changes (mutations) in DNA when he realized that he had instead invented a method of amplifying any

DNA region through repeated cycles of duplication driven by DNA polymerase. In *Scientific American*, Mullis summarized the procedure: "Beginning with a single molecule of the genetic material DNA, the PCR can generate 100 billion similar molecules in an afternoon. The reaction is easy to execute. It requires no more than a test tube, a few simple reagents, and a source of heat." He was awarded the Nobel Prize in Chemistry in 1993 for his invention, seven years after he and his colleagues at Cetus first put his proposal to practice. However, some controversies have remained about the intellectual and practical contributions of other scientists to Mullis' work, and whether he had been the sole inventor of the PCR principle.

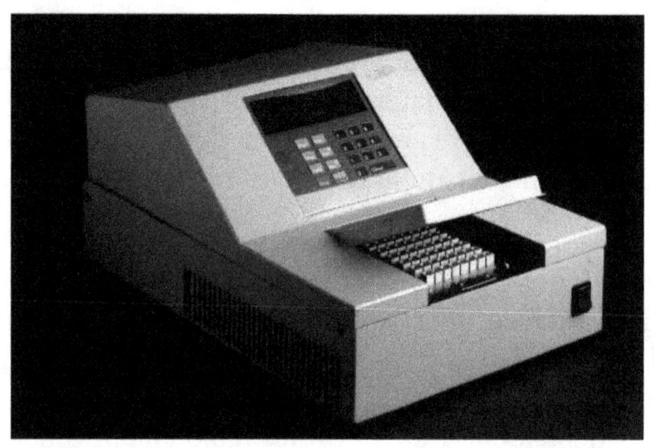

"Baby Blue", a 1986 prototype machine for doing PCR

At the core of the PCR method is the use of a suitable DNA polymerase able to withstand the high temperatures of >90 °C (194 °F) required for separation of the two DNA strands in the DNA double helix after each replication cycle. The DNA polymerases initially employed for in vitro experiments presaging PCR were unable to withstand these high temperatures. So the early procedures for DNA replication were very inefficient and time-consuming, and required large amounts of DNA polymerase and continuous handling throughout the process.

The discovery in 1976 of Taq polymerase — a DNA polymerase purified from the thermophilic bacterium, *Thermus aquaticus*, which naturally lives in hot (50 to 80 °C (122 to 176 °F)) environments such as hot springs — paved the way for dramatic improvements of the PCR method. The DNA polymerase isolated from *T. aquaticus* is stable at high temperatures remaining active even after DNA denaturation, thus obviating the need to add new DNA polymerase after each cycle. This allowed an automated thermocycler-based process for DNA amplification.

Patent Disputes

The PCR technique was patented by Kary Mullis and assigned to Cetus Corporation, where Mullis worked when he invented the technique in 1983. The *Taq* polymerase enzyme was also covered by patents. There have been several high-profile lawsuits related to the technique, including an unsuccessful lawsuit brought by DuPont. The pharmaceutical company Hoffmann-La Roche purchased the rights to the patents in 1992 and currently holds those that are still protected.

A related patent battle over the Taq polymerase enzyme is still ongoing in several jurisdictions around the world between Roche and Promega. The legal arguments have extended beyond the lives of the original PCR and Taq polymerase patents, which expired on March 28, 2005.

Gel Electrophoresis

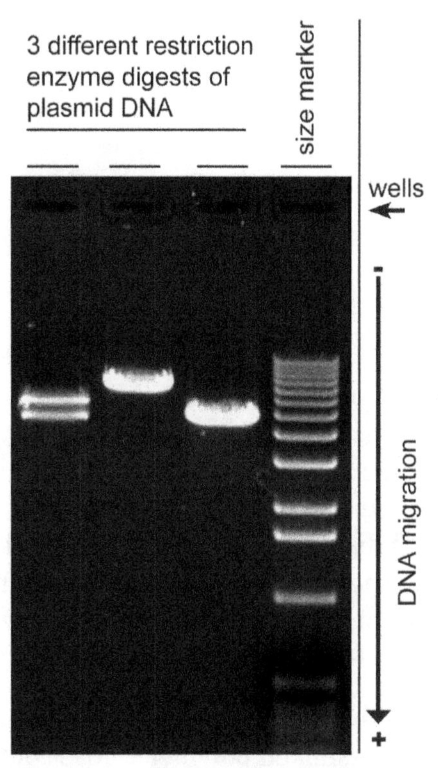

Digital image of 3 plasmid restriction digests run on a 1% w/v agarose gel, 3 volt/cm, stained with ethidium bromide. The DNA size marker is a commercial 1 kbp ladder. The position of the wells and direction of DNA migration is noted.

Gel electrophoresis is a method for separation and analysis of macromolecules (DNA, RNA and proteins) and their fragments, based on their size and charge. It is used in clinical chemistry to separate proteins by charge and/or size (IEF agarose, essentially size independent) and in biochemistry and molecular biology to separate a mixed population of DNA and RNA fragments by length, to estimate the size of DNA and RNA fragments or to separate proteins by charge.

Nucleic acid molecules are separated by applying an electric field to move the negatively charged molecules through a matrix of agarose or other substances. Shorter molecules move faster and migrate farther than longer ones because shorter molecules migrate more easily through the pores of the gel. This phenomenon is called sieving. Proteins are separated by charge in agarose because the pores of the gel are too large to sieve proteins. Gel electrophoresis can also be used for separation of nanoparticles.

Gel electrophoresis uses a gel as an anticonvective medium and/or sieving medium during electrophoresis, the movement of a charged particle in an electrical field. Gels suppress the thermal convection caused by application of the electric field, and can also act as a sieving medium, retarding the passage of molecules; gels can also simply serve to maintain the finished separation, so that a post electrophoresis stain can be applied. DNA Gel electrophoresis is usually performed for analytical purposes, often after amplification of DNA via PCR, but may be used as a preparative technique prior to use of other methods such as mass spectrometry, RFLP, PCR, cloning, DNA sequencing, or Southern blotting for further characterization.

Physical Basis

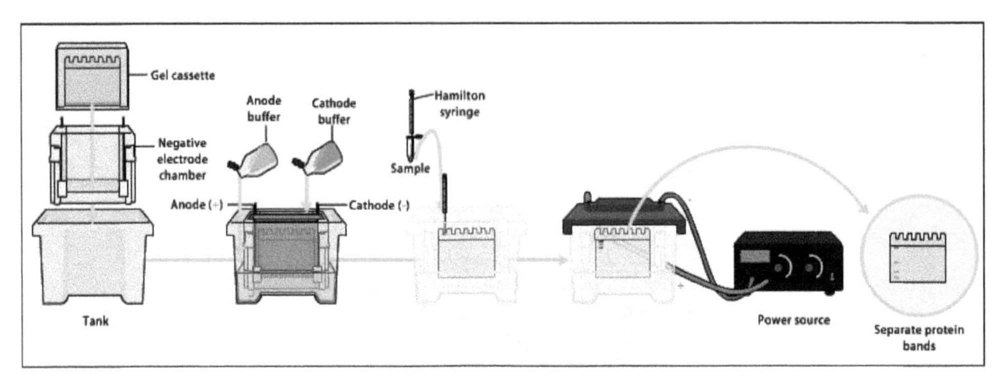

Overview of Gel Electrophoresis.

In simple terms, electrophoresis is a process which enables the sorting of molecules based on size. Using an electric field, molecules (such as DNA) can be made to move through a gel made of agar or polyacrylamide. The electric field consists of a negative charge at one end which pushes the molecules through the gel, and a positive charge at the other end that pulls the molecules through the gel. The molecules being sorted are dispensed into a well in the gel material. The gel is placed in an electrophoresis chamber, which is then connected to a power source. When the electric current is applied, the larger molecules move more slowly through the gel while the smaller molecules move faster. The different sized molecules form distinct bands on the gel.

The term "gel" in this instance refers to the matrix used to contain, then separate the target molecules. In most cases, the gel is a crosslinked polymer whose composition and porosity is chosen based on the specific weight and composition of the target to be analyzed. When separating proteins or small nucleic acids (DNA, RNA, or oligonucleotides) the gel is usually composed of different concentrations of acrylamide and a cross-linker, producing different sized mesh networks of polyacrylamide. When separating larger nucleic acids (greater than a few hundred bases), the preferred matrix is purified agarose. In both cases, the gel forms a solid, yet porous matrix. Acrylamide, in contrast to polyacrylamide, is a neurotoxin and must be handled using appropriate safety precautions to avoid poisoning. Agarose is composed of long unbranched chains of uncharged carbohydrate without cross links resulting in a gel with large pores allowing for the separation of macromolecules and macromolecular complexes.

"Electrophoresis" refers to the electromotive force (EMF) that is used to move the molecules through the gel matrix. By placing the molecules in wells in the gel and applying an electric field, the molecules will move through the matrix at different rates, determined largely by their mass when the charge to mass ratio (Z) of all species is uniform. However, when charges are not all uniform then, the electrical field generated by the electrophoresis procedure will affect the species that have different charges and therefore will attract the species according to their charges being the opposite. Species that are positively charged will migrate towards the cathode which is negatively charged (because this is an electrolytic rather than galvanic cell). If the species are negatively charged they will migrate towards the positively charged anode.

If several samples have been loaded into adjacent wells in the gel, they will run parallel in individual lanes. Depending on the number of different molecules, each lane shows separation of the components from the original mixture as one or more distinct bands, one band per component.

Incomplete separation of the components can lead to overlapping bands, or to indistinguishable smears representing multiple unresolved components. Bands in different lanes that end up at the same distance from the top contain molecules that passed through the gel with the same speed, which usually means they are approximately the same size. There are molecular weight size markers available that contain a mixture of molecules of known sizes. If such a marker was run on one lane in the gel parallel to the unknown samples, the bands observed can be compared to those of the unknown in order to determine their size. The distance a band travels is approximately inversely proportional to the logarithm of the size of the molecule.

There are limits to electrophoretic techniques. Since passing current through a gel causes heating, gels may melt during electrophoresis. Electrophoresis is performed in buffer solutions to reduce pH changes due to the electric field, which is important because the charge of DNA and RNA depends on pH, but running for too long can exhaust the buffering capacity of the solution. Further, different preparations of genetic material may not migrate consistently with each other, for morphological or other reasons.

Types of Gel

The types of gel most typically used are agarose and polyacrylamide gels. Each type of gel is well-suited to different types and sizes of analyte. Polyacrylamide gels are usually used for proteins, and have very high resolving power for small fragments of DNA (5-500 bp). Agarose gels on the other hand have lower resolving power for DNA but have greater range of separation, and are therefore used for DNA fragments of usually 50-20,000 bp in size, but resolution of over 6 Mb is possible with pulsed field gel electrophoresis (PFGE). Polyacrylamide gels are run in a vertical configuration while agarose gels are typically run horizontally in a submarine mode. They also differ in their casting methodology, as agarose sets thermally, while polyacrylamide forms in a chemical polymerization reaction.

Agarose

Agarose gels are made from the natural polysaccharide polymers extracted from seaweed. Agarose gels are easily cast and handled compared to other matrices, because the gel setting is a physical rather than chemical change. Samples are also easily recovered. After the experiment is finished, the resulting gel can be stored in a plastic bag in a refrigerator.

Agarose gels do not have a uniform pore size, but are optimal for electrophoresis of proteins that are larger than 200 kDa. Agarose gel electrophoresis can also be used for the separation of DNA fragments ranging from 50 base pair to several megabases (millions of bases), the largest of which require specialized apparatus. The distance between DNA bands of different lengths is influenced by the percent agarose in the gel, with higher percentages requiring longer run times, sometimes days. Instead high percentage agarose gels should be run with a pulsed field electrophoresis (PFE), or field inversion electrophoresis.

"Most agarose gels are made with between 0.7% (good separation or resolution of large 5–10kb DNA fragments) and 2% (good resolution for small 0.2–1kb fragments) agarose dissolved in electrophoresis buffer. Up to 3% can be used for separating very tiny fragments but a vertical polyacrylamide gel is more appropriate in this case. Low percentage gels are very weak and may break when you try to lift them. High percentage gels are often brittle and do not set evenly. 1% gels are common for many applications."

Polyacrylamide

Polyacrylamide gel electrophoresis (PAGE) is used for separating proteins ranging in size from 5 to 2,000 kDa due to the uniform pore size provided by the polyacrylamide gel. Pore size is controlled by modulating the concentrations of acrylamide and bis-acrylamide powder used in creating a gel. Care must be used when creating this type of gel, as acrylamide is a potent neurotoxin in its liquid and powdered forms.

Traditional DNA sequencing techniques such as Maxam-Gilbert or Sanger methods used polyacrylamide gels to separate DNA fragments differing by a single base-pair in length so the sequence could be read. Most modern DNA separation methods now use agarose gels, except for particularly small DNA fragments. It is currently most often used in the field of immunology and protein analysis, often used to separate different proteins or isoforms of the same protein into separate bands. These can be transferred onto a nitrocellulose or PVDF membrane to be probed with antibodies and corresponding markers, such as in a western blot.

Typically resolving gels are made in 6%, 8%, 10%, 12% or 15%. Stacking gel (5%) is poured on top of the resolving gel and a gel comb (which forms the wells and defines the lanes where proteins, sample buffer and ladders will be placed) is inserted. The percentage chosen depends on the size of the protein that one wishes to identify or probe in the sample. The smaller the known weight, the higher the percentage that should be used. Changes on the buffer system of the gel can help to further resolve proteins of very small sizes.

Starch

Partially hydrolysed potato starch makes for another non-toxic medium for protein electrophoresis. The gels are slightly more opaque than acrylamide or agarose. Non-denatured proteins can be separated according to charge and size. They are visualised using Napthal Black or Amido Black staining. Typical starch gel concentrations are 5% to 10%.

Gel Conditions

Denaturing

TTGE profiles representing the bifidobacterial diversity of fecal samples from two healthy volunteers (A and B) before and after AMC (Oral Amoxicillin-Clavulanic Acid) treatment

Denaturing gels are run under conditions that disrupt the natural structure of the analyte, causing it to unfold into a linear chain. Thus, the mobility of each macromolecule depends only on its linear length and its mass-to-charge ratio. Thus, the secondary, tertiary, and quaternary levels of biomolecular structure are disrupted, leaving only the primary structure to be analyzed.

Nucleic acids are often denatured by including urea in the buffer, while proteins are denatured using sodium dodecyl sulfate, usually as part of the SDS-PAGE process. For full denaturation of proteins, it is also necessary to reduce the covalent disulfide bonds that stabilize their tertiary and quaternary structure, a method called reducing PAGE. Reducing conditions are usually maintained by the addition of beta-mercaptoethanol or dithiothreitol. For general analysis of protein samples, reducing PAGE is the most common form of protein electrophoresis.

Denaturing conditions are necessary for proper estimation of molecular weight of RNA. RNA is able to form more intramolecular interactions than DNA which may result in change of its electrophoretic mobility. Urea, DMSO and glyoxal are the most often used denaturing agents to disrupt RNA structure. Originally, highly toxic methylmercury hydroxide was often used in denaturing RNA electrophoresis, but it may be method of choice for some samples.

Denaturing gel electrophoresis is used in the DNA and RNA banding pattern-based methods DGGE (denaturing gradient gel electrophoresis), TGGE (temperature gradient gel electrophoresis), and TTGE (temporal temperature gradient electrophoresis).

Native

Specific enzyme-linked staining: Glucose-6-Phosphate Dehydrogenase isoenzymes in *Plasmodium falciparum* infected Red blood cells

Native gels are run in non-denaturing conditions, so that the analyte's natural structure is maintained. This allows the physical size of the folded or assembled complex to affect the mobility, allowing for analysis of all four levels of the biomolecular structure. For biological samples, detergents are used only to the extent that they are necessary to lyse lipid membranes in the cell. Complexes remain—for the most part—associated and folded as they would be in the cell. One downside, however, is that complexes may not separate cleanly or predictably, as it is difficult to predict how the molecule's shape and size will affect its mobility. Addressing and solving this problem is a major aim of quantitative native PAGE.

Unlike denaturing methods, native gel electrophoresis does not use a charged denaturing agent. The molecules being separated (usually proteins or nucleic acids) therefore differ not only in molecular mass and intrinsic charge, but also the cross-sectional area, and thus experience different electrophoretic forces dependent on the shape of the overall structure. For proteins, since they remain in the native state they may be visualised not only by general protein staining reagents but also by specific enzyme-linked staining.

Native gel electrophoresis is typically used in proteomics and metallomics. However, native PAGE is also used to scan genes (DNA) for unknown mutations as in Single-strand conformation polymorphism.

Buffers

Buffers in gel electrophoresis are used to provide ions that carry a current and to maintain the pH at a relatively constant value. There are a number of buffers used for electrophoresis. The most common being, for nucleic acids Tris/Acetate/EDTA (TAE), Tris/Borate/EDTA (TBE). Many other buffers have been proposed, e.g. lithium borate, which is almost never used, based on Pubmed citations (LB), iso electric histidine, pK matched goods buffers, etc.; in most cases the purported rationale is lower current (less heat) and or matched ion mobilities, which leads to longer buffer life. Borate is problematic; Borate can polymerize, and/or interact with cis diols such as those found in RNA. TAE has the lowest buffering capacity but provides the best resolution for larger DNA. This means a lower voltage and more time, but a better product. LB is relatively new and is ineffective in resolving fragments larger than 5 kbp; However, with its low conductivity, a much higher voltage could be used (up to 35 V/cm), which means a shorter analysis time for routine electrophoresis. As low as one base pair size difference could be resolved in 3% agarose gel with an extremely low conductivity medium (1 mM Lithium borate).

Most SDS-PAGE protein separations are performed using a "discontinuous" (or DISC) buffer system that significantly enhances the sharpness of the bands within the gel. During electrophoresis in a discontinuous gel system, an ion gradient is formed in the early stage of electrophoresis that causes all of the proteins to focus into a single sharp band in a process called isotachophoresis. Separation of the proteins by size is achieved in the lower, "resolving" region of the gel. The resolving gel typically has a much smaller pore size, which leads to a sieving effect that now determines the electrophoretic mobility of the proteins.

Visualization

After the electrophoresis is complete, the molecules in the gel can be stained to make them visible. DNA may be visualized using ethidium bromide which, when intercalated into DNA, fluoresce under ultraviolet light, while protein may be visualised using silver stain or Coomassie Brilliant Blue dye. Other methods may also be used to visualize the separation of the mixture's components on the gel. If the molecules to be separated contain radioactivity, for example in a DNA sequencing gel, an autoradiogram can be recorded of the gel. Photographs can be taken of gels, often using a Gel Doc system.

Downstream Processing

After separation, an additional separation method may then be used, such as isoelectric focusing or SDS-PAGE. The gel will then be physically cut, and the protein complexes extracted from each portion separately. Each extract may then be analysed, such as by peptide mass fingerprinting or de novo peptide sequencing after in-gel digestion. This can provide a great deal of information about the identities of the proteins in a complex.

Applications

- Estimation of the size of DNA molecules following restriction enzyme digestion, e.g. in restriction mapping of cloned DNA.
- Analysis of PCR products, e.g. in molecular genetic diagnosis or genetic fingerprinting
- Separation of restricted genomic DNA prior to Southern transfer, or of RNA prior to Northern transfer.

Gel electrophoresis is used in forensics, molecular biology, genetics, microbiology and biochemistry. The results can be analyzed quantitatively by visualizing the gel with UV light and a gel imaging device. The image is recorded with a computer operated camera, and the intensity of the band or spot of interest is measured and compared against standard or markers loaded on the same gel. The measurement and analysis are mostly done with specialized software.

Depending on the type of analysis being performed, other techniques are often implemented in conjunction with the results of gel electrophoresis, providing a wide range of field-specific applications.

Nucleic Acids

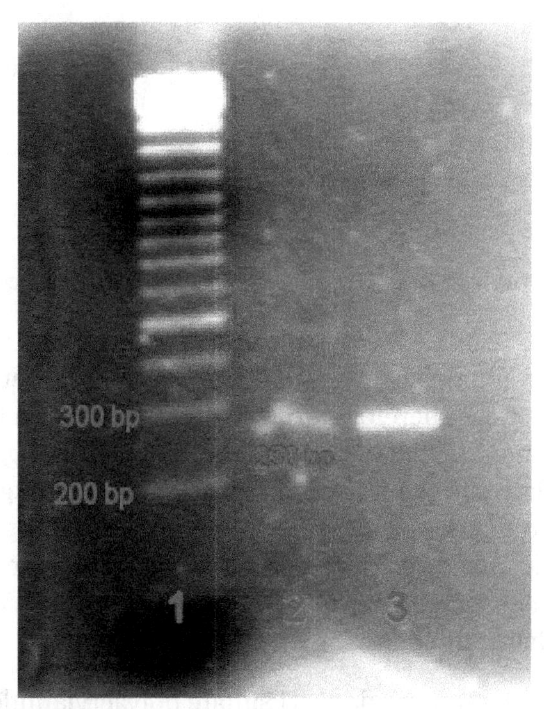

An agarose gel of a PCR product compared to a DNA ladder.

In the case of nucleic acids, the direction of migration, from negative to positive electrodes, is due to the naturally occurring negative charge carried by their sugar-phosphate backbone.

Double-stranded DNA fragments naturally behave as long rods, so their migration through the gel is relative to their size or, for cyclic fragments, their radius of gyration. Circular DNA such as plasmids, however, may show multiple bands, the speed of migration may depend on whether it is relaxed or supercoiled. Single-stranded DNA or RNA tend to fold up into molecules with complex shapes and migrate through the gel in a complicated manner based on their tertiary structure. Therefore, agents that disrupt the hydrogen bonds, such as sodium hydroxide or formamide, are used to denature the nucleic acids and cause them to behave as long rods again.

Gel electrophoresis of large DNA or RNA is usually done by agarose gel electrophoresis. See the "Chain termination method" page for an example of a polyacrylamide DNA sequencing gel. Characterization through ligand interaction of nucleic acids or fragments may be performed by mobility shift affinity electrophoresis.

Electrophoresis of RNA samples can be used to check for genomic DNA contamination and also for RNA degradation. RNA from eukaryotic organisms shows distinct bands of 28s and 18s rRNA, the 28s band being approximately twice as intense as the 18s band. Degraded RNA has less sharply defined bands, has a smeared appearance, and intensity ratio is less than 2:1.

Proteins

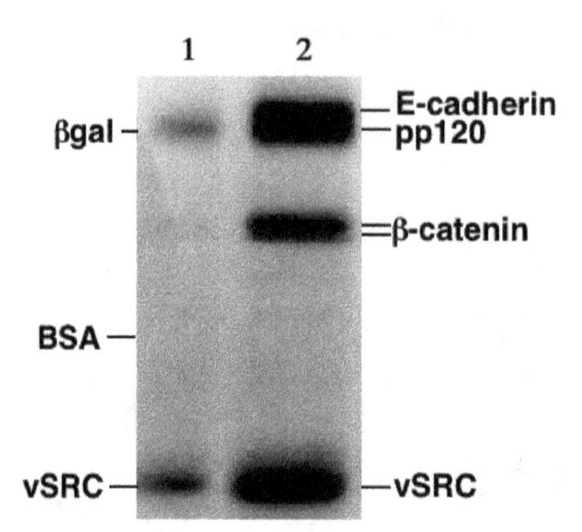

SDS-PAGE autoradiography – The indicated proteins are present in different concentrations in the two samples.

Proteins, unlike nucleic acids, can have varying charges and complex shapes, therefore they may not migrate into the polyacrylamide gel at similar rates, or at all, when placing a negative to positive EMF on the sample. Proteins therefore, are usually denatured in the presence of a detergent such as sodium dodecyl sulfate (SDS) that coats the proteins with a negative charge. Generally, the amount of SDS bound is relative to the size of the protein (usually 1.4g SDS per gram of protein), so that the resulting denatured proteins have an overall negative charge, and all the proteins have a similar charge to mass ratio. Since denatured proteins act like long rods instead of having a complex tertiary shape, the rate at which the resulting SDS coated proteins migrate in the gel is relative only to its size and not its charge or shape.

Proteins are usually analyzed by sodium dodecyl sulfate polyacrylamide gel electrophoresis (SDS-PAGE), by native gel electrophoresis, by preparative gel electrophoresis (QPNC-PAGE), or by 2-D electrophoresis.

Characterization through ligand interaction may be performed by electroblotting or by affinity electrophoresis in agarose or by capillary electrophoresis as for estimation of binding constants and determination of structural features like glycan content through lectin binding.

History

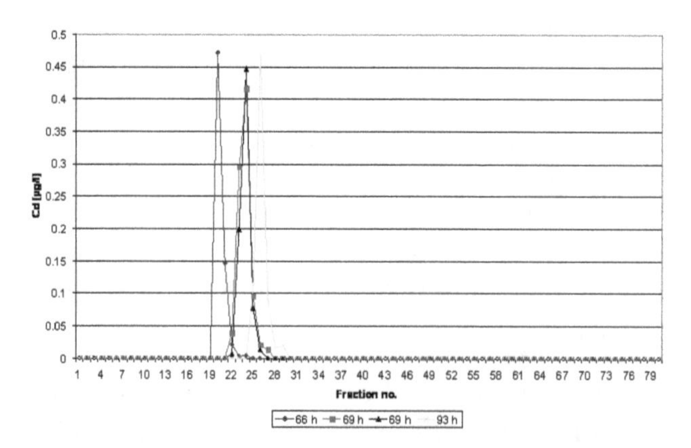

Electropherogram showing four PAGE runs of a native chromatographically prepurified high molecular weight cadmium protein (≈ 200 ku) as a function of time of polymerization of the gel (4% T, 2.67% C).

- 1930s – first reports of the use of sucrose for gel electrophoresis
- 1955 – introduction of starch gels, mediocre separation (Smithies)
- 1959 – introduction of acrylamide gels; disc electrophoresis (Ornstein and Davis); accurate control of parameters such as pore size and stability; and (Raymond and Weintraub)
- 1966 – first use of agar gels
- 1969 – introduction of denaturing agents especially SDS separation of protein subunit (Weber and Osborn)
- 1970 – Laemmli separated 28 components of T4 phage using a stacking gel and SDS
- 1972 – agarose gels with ethidium bromide stain
- 1975 – 2-dimensional gels (O'Farrell); isoelectric focusing then SDS gel electrophoresis
- 1977 – sequencing gels
- 1983 – pulsed field gel electrophoresis enables separation of large DNA molecules
- 1983 – introduction of capillary electrophoresis
- 2004 – standardized time of polymerization of acrylamide gels enables clean and predictable separation of native proteins (Kastenholz)

A 1959 book on electrophoresis by Milan Bier cites references from the 1800s. However, Oliver Smithies made significant contributions. Bier states: "The method of Smithies ... is finding wide application because of its unique separatory power." Taken in context, Bier clearly implies that Smithies' method is an improvement.

Polymer Used in Gel Electrophoresis

Polyacrylamide

Polyacrylamide (IUPAC poly(2-propenamide) or poly(1-carbamoylethylene), abbreviated as PAM) is a polymer ($-CH_2CHCONH_2-$) formed from acrylamide subunits. It can be synthesized as a simple linear-chain structure or cross-linked, typically using *N,N'*-methylenebisacrylamide. In the cross-linked form, the possibility of the monomer being present is reduced even further. It is highly water-absorbent, forming a soft gel when hydrated, used in such applications as polyacrylamide gel electrophoresis and in manufacturing soft contact lenses. In the straight-chain form, it is also used as a thickener and suspending agent. More recently, it has been used as a subdermal filler for aesthetic facial surgery.

Uses of Polyacrylamide

One of the largest uses for polyacrylamide is to flocculate solids in a liquid. This process applies to water treatment, and processes like paper making and screen printing. Polyacrylamide can be supplied in a powder or liquid form, with the liquid form being subcategorized as solution and emulsion polymer. Even though these products are often called 'polyacrylamide', many are actually copolymers of acrylamide and one or more other chemical species, such as an acrylic acid or a salt thereof. The main consequence of this is to give the 'modified' polymer a particular ionic character.

Another common use of polyacrylamide and its derivatives is in subsurface applications such as Enhanced Oil Recovery. High viscosity aqueous solutions can be generated with low concentrations of polyacrylamide polymers, and these can be injected to improve the economics of conventional waterflooding.

It has also been advertised as a soil conditioner called Krilium by Monsanto Company in the 1950s and today "MP", which is stated to be a "unique formulation of PAM (water-soluble polyacrylamide)". It is often used for horticultural and agricultural use under trade names such as Broadleaf P4, Swell-Gel and so on. The anionic form of cross-linked polyacrylamide is frequently used as a soil conditioner on farm land and construction sites for erosion control, in order to protect the water quality of nearby rivers and streams.

The polymer is also used to make Gro-Beast toys, which expand when placed in water, such as the Test Tube Aliens. Similarly, the absorbent properties of one of its copolymers can be utilized as an additive in body-powder.

The ionic form of polyacrylamide has found an important role in the potable water treatment industry. Trivalent metal salts like ferric chloride and aluminum chloride are bridged by the long polymer chains of polyacrylamide. This results in significant enhancement of the flocculation rate. This allows water treatment plants to greatly improve the removal of total organic content (TOC) from raw water.

Polyacrylamide is often used in molecular biology applications as a medium for electrophoresis of proteins and nucleic acids in a technique known as PAGE.

It was also used in the synthesis of the first Boger fluid.

Soil Conditioner

The primary functions of Polyacrylamide Soil Conditioners are to increase soil tilth, aeration, and porosity and reduce compaction, dustiness and water run-off. Secondary functions are to increase plant vigor, color, appearance, rooting depth and emergence of seeds while decreasing water requirements, diseases, erosion and maintenance expenses. FC 2712 is used for this purpose.

Stability

In dilute aqueous solution, such as is commonly used for Enhanced Oil Recovery applications, polyacrylamide polymers are susceptible to chemical, thermal, and mechanical degradation. Chemical degradation occurs when the labile amine moiety hydrolyzes at elevated temperature or pH, resulting in the evolution of ammonia and a remaining carboxyl group. Thus, the degree of anionicity of the molecule increases. Thermal degradation of the vinyl backbone can occur through several possible radical mechanisms, including the autooxidation of small amounts of iron and reactions between oxygen and residual impurities from polymerization at elevated temperature. Mechanical degradation can also be an issue at the high shear rates experienced in the near-wellbore region.

Environmental Effects

Concerns have been raised that polyacrylamide used in agriculture may contaminate food with acrylamide, a known neurotoxin. While polyacrylamide itself is relatively non-toxic, it is known that commercially available polyacrylamide contains minute residual amounts of acrylamide remaining from its production, usually less than 0.05% w/w.

Additionally, there are concerns that polyacrylamide may de-polymerise to form acrylamide. In a study conducted in 2003 at the Central Science Laboratory in Sand Hutton, England, polyacrylamide was treated similarly as food during cooking. It was shown that these conditions do not cause polyacrylamide to de-polymerise significantly.

In a study conducted in 1997 at Kansas State University, the effect of environmental conditions on polyacrylamide were tested, and it was shown that degradation of polyacrylamide under certain conditions can cause the release of acrylamide. The experimental design of this study as well as its results and their interpretation have been questioned, and a 1999 study by the Nalco Chemical Company did not replicate the results.

DNA Sequencing

DNA sequencing is the process of determining the precise order of nucleotides within a DNA molecule. It includes any method or technology that is used to determine the order of the four bases—adenine, guanine, cytosine, and thymine—in a strand of DNA. The advent of rapid DNA sequencing methods has greatly accelerated biological and medical research and discovery.

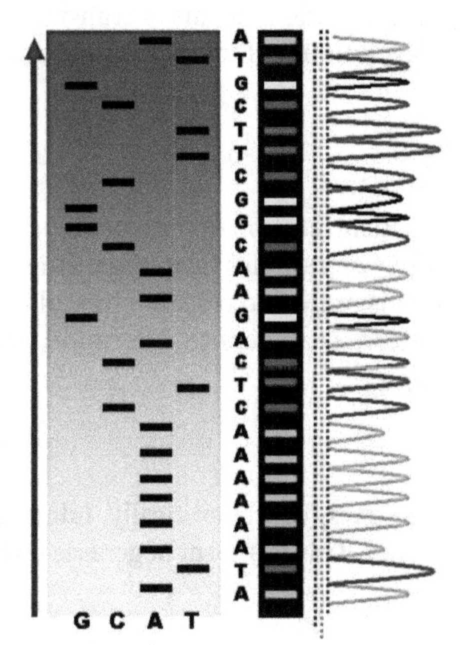

An example of the results of automated chain-termination DNA sequencing.

The first DNA sequences were obtained in the early 1970s by academic researchers using laborious methods based on two-dimensional chromatography. Following the development of fluorescence-based sequencing methods with a DNA sequencer, DNA sequencing has become easier and orders of magnitude faster.

Knowledge of DNA sequences has become indispensable for basic biological research, and in numerous applied fields such as medical diagnosis, biotechnology, forensic biology, virology and biological systematics. The rapid speed of sequencing attained with modern DNA sequencing technology has been instrumental in the sequencing of complete DNA sequences, or genomes of numerous types and species of life, including the human genome and other complete DNA sequences of many animal, plant, and microbial species.

Applications

DNA sequencing may be used to determine the sequence of individual genes, larger genetic regions (i.e. clusters of genes or operons), full chromosomes or entire genomes, of any organism. DNA sequencing is also the most efficient way to sequence RNA or proteins (via their open reading frames). In fact, DNA sequencing has become a key technology in many areas of biology and other sciences such as medicine, forensics, or anthropology. Some applications include:

Molecular Biology

Sequencing is used in molecular biology to study genomes and the proteins they encode. Information obtained using sequencing allows researchers to identify changes in genes, associations with diseases and phenotypes, and identify potential drug targets.

Evolutionary Biology

Since DNA is an informative macromolecule in terms of transmission from one generation to another, DNA sequencing is used in evolutionary biology to study how different organisms are related and how they evolved.

Metagenomics

The field of metagenomics involves identification of organisms present in a body of water, sewage, dirt, debris filtered from the air, or swab samples from organisms. Knowing which organisms are present in a particular environment is critical to research in ecology, epidemiology, microbiology, and other fields. Sequencing enables researchers to determine which types of microbes may be present in a microbiome, for example.

Medicine

Medical technicians may sequence genes (or, theoretically, full genomes) from patients to determine if there is risk of genetic diseases. This is a form of genetic testing, though some genetic tests may not involve DNA sequencing.

Forensics

DNA sequencing may be used along with DNA profiling methods for forensic identification and paternity testing.

The Four Canonical Bases

The canonical structure of DNA has four bases: thymine (T), adenine (A), cytosine (C), and guanine (G). DNA sequencing is the determination of the physical order of these bases in a molecule of DNA. However, there are many other bases that may be present in a molecule. In some viruses (specifically, bacteriophage), cytosine may be replaced by hydroxy methyl or hydroxy methyl glucose cytosine. In mammalian DNA, variant bases with methyl groups or phosphosulfate may be found. Depending on the sequencing technique, a particular modification, e.g., the 5mC (5 methyl cytosine) common in humans, may or may not be detected.

History

Deoxyribonucleic acid (DNA) was first discovered and isolated by Friedrich Miescher in 1869, but it remained understudied for many decades because proteins, rather than DNA, were thought to hold the genetic blueprint to life. This situation changed after 1944 as a result of some experiments by Oswald Avery, Colin MacLeod, and Maclyn McCarty demonstrating that purified DNA could change one strain of bacteria into another. This was the first time that DNA was shown capable of transforming the properties of cells.

In 1953 James Watson and Francis Crick put forward their double-helix model of DNA, based on crystallized X-ray structures being studied by Rosalind Franklin. According to the model, DNA is composed of two strands of nucleotides coiled around each other, linked together by hydrogen bonds and running in opposite directions. Each strand is composed of four complementary nucleotides – adenine (A), cytosine (C), guanine (G) and thymine (T) – with an A on one strand always paired with T on the other, and C always paired with G. They proposed such a structure allowed each strand to be used to reconstruct the other, an idea central to the passing on of hereditary information between generations.

Frederick Sanger, a pioneer of sequencing. Sanger is one of the few scientists who was awarded two Nobel prizes, one for the sequencing of proteins, and the other for the sequencing of DNA.

The foundation for sequencing proteins was first laid by the work of Fred Sanger who by 1955 had completed the sequence of all the amino acids in insulin, a small protein secreted by the pancreas. This provided the first conclusive evidence that proteins were chemical entities with a specific molecular pattern rather than a random mixture of material suspended in fluid. Sanger's success in sequencing insulin greatly electrified x-ray crystallographers, including Watson and Crick who by now were trying to understand how DNA directed the formation of proteins within a cell. Soon after attending a series of lectures given by Fred Sanger in October 1954, Crick began to develop a theory which argued that the arrangement of nucleotides in DNA determined the sequence of amino acids in proteins which in turn helped determine the function of a protein. He published this theory in 1958.

RNA Sequencing

RNA sequencing was one of the earliest forms of nucleotide sequencing. The major landmark of RNA sequencing is the sequence of the first complete gene and the complete genome of Bacteriophage MS2, identified and published by Walter Fiers and his coworkers at the University of Ghent (Ghent, Belgium), in 1972 and 1976. Traditional RNA sequencing methods require the creation of a cDNA molecule which must be sequenced. In August 2016, scientists at Oxford Nanopore Technologies published a pre-print indicating that direct RNA sequencing could be performed in real time using nanopore technology.

Early DNA Sequencing Methods

The first method for determining DNA sequences involved a location-specific primer extension strategy established by Ray Wu at Cornell University in 1970. DNA polymerase catalysis and specific nucleotide labeling, both of which figure prominently in current sequencing schemes, were used to sequence the cohesive ends of lambda phage DNA. Between 1970 and 1973, Wu, R Padmanabhan and colleagues demonstrated that this method can be employed to determine any DNA sequence using synthetic location-specific primers. Frederick Sanger then adopted this primer-extension strategy to develop more rapid DNA sequencing methods at the MRC Centre, Cambridge, UK and published a method for "DNA sequencing with chain-terminating inhibitors" in 1977. Walter Gilbert and Allan Maxam at Harvard also developed sequencing methods, including one for "DNA sequencing by chemical degradation". In 1973, Gilbert and Maxam reported the sequence of 24 basepairs using a method known as wandering-spot analysis. Advancements in sequencing were aided by the concurrent development of recombinant DNA technology, allowing DNA samples to be isolated from sources other than viruses.

Sequencing of Full Genomes

The first full DNA genome to be sequenced was that of bacteriophage φX174 in 1977. Medical Research Council scientists deciphered the complete DNA sequence of the Epstein-Barr virus in 1984, finding it contained 172,282 nucleotides. Completion of the sequence marked a significant turning point in DNA sequencing because it was achieved with no prior genetic profile knowledge of the virus.

A non-radioactive method for transferring the DNA molecules of sequencing reaction mixtures onto an immobilizing matrix during electrophoresis was developed by Pohl and co-workers in the early 1980s. Followed by the commercialization of the DNA sequencer "Direct-Blotting-Electrophoresis-System GATC 1500" by GATC Biotech, which was intensively used in the framework of the EU genome-sequencing programme, the complete DNA sequence of the yeast *Saccharomyces cerevisiae* chromosome II. Leroy E. Hood's laboratory at the California Institute of Technology announced the first semi-automated DNA sequencing machine in 1986. This was followed by Applied Biosystems' marketing of the first fully automated sequencing machine, the ABI 370, in 1987 and by Dupont's Genesis 2000 which used a novel fluorescent labeling technique enabling all four dideoxynucleotides to be identified in a single lane. By 1990, the U.S. National Institutes of Health (NIH) had begun large-scale sequencing trials on *Mycoplasma capricolum, Escherichia coli, Caenorhabditis elegans,* and *Saccharomyces cerevisiae* at a cost of US$0.75 per base. Meanwhile, sequencing of human cDNA sequences called expressed sequence tags began in Craig Venter's lab, an attempt to capture the coding fraction of the human genome. In 1995, Venter, Hamilton Smith, and colleagues at The Institute for Genomic Research (TIGR) published the first complete genome of a free-living organism, the bacterium *Haemophilus influenzae.* The circular chromosome contains 1,830,137 bases and its publication in the journal Science marked the first published use of whole-genome shotgun sequencing, eliminating the need for initial mapping efforts.

By 2001, shotgun sequencing methods had been used to produce a draft sequence of the human genome.

Next-generation Sequencing (NGS) Methods

Several new methods for DNA sequencing were developed in the mid to late 1990s and were implemented in commercial DNA sequencers by the year 2000.

On October 26, 1990, Roger Tsien, Pepi Ross, Margaret Fahnestock and Allan J Johnston filed a patent describing stepwise ("base-by-base") sequencing with removable 3' blockers on DNA arrays (blots and single DNA molecules). In 1996, Pål Nyrén and his student Mostafa Ronaghi at the Royal Institute of Technology in Stockholm published their method of pyrosequencing.

On April 1, 1997, Pascal Mayer and Laurent Farinelli submitted patents to the World Intellectual Property Organization describing DNA colony sequencing. The DNA sample preparation and random surface-PCR arraying methods described in this patent, coupled to Roger Tsien et al.'s "base-by-base" sequencing method, is now implemented in Illumina's Hi-Seq genome sequencers.

Lynx Therapeutics published and marketed "Massively parallel signature sequencing", or MPSS, in 2000. This method incorporated a parallelized, adapter/ligation-mediated, bead-based sequencing technology and served as the first commercially available "next-generation" sequencing method, though no DNA sequencers were sold to independent laboratories.

In 2004, 454 Life Sciences marketed a parallelized version of pyrosequencing. The first version of their machine reduced sequencing costs 6-fold compared to automated Sanger sequencing, and was the second of the new generation of sequencing technologies, after MPSS.

The large quantities of data produced by DNA sequencing have also required development of new methods and programs for sequence analysis. Phil Green and Brent Ewing of the University of Washington described their phred quality score for sequencer data analysis in 1998.

Basic Methods

Maxam-Gilbert Sequencing

Allan Maxam and Walter Gilbert published a DNA sequencing method in 1977 based on chemical modification of DNA and subsequent cleavage at specific bases. Also known as chemical sequencing, this method allowed purified samples of double-stranded DNA to be used without further cloning. This method's use of radioactive labeling and its technical complexity discouraged extensive use after refinements in the Sanger methods had been made.

Maxam-Gilbert sequencing requires radioactive labeling at one 5' end of the DNA and purification of the DNA fragment to be sequenced. Chemical treatment then generates breaks at a small proportion of one or two of the four nucleotide bases in each of four reactions (G, A+G, C, C+T). The concentration of the modifying chemicals is controlled to introduce on average one modification per DNA molecule. Thus a series of labeled fragments is generated, from the radiolabeled end to the first "cut" site in each molecule. The fragments in the four reactions are electrophoresed side by side in denaturing acrylamide gels for size separation. To visualize the fragments, the gel is exposed to X-ray film for autoradiography, yielding a series of dark bands each corresponding to a radiolabeled DNA fragment, from which the sequence may be inferred.

Chain-termination Methods

The chain-termination method developed by Frederick Sanger and coworkers in 1977 soon became the method of choice, owing to its relative ease and reliability. When invented, the chain-terminator method used fewer toxic chemicals and lower amounts of radioactivity than the Maxam and Gilbert method. Because of its comparative ease, the Sanger method was soon automated and was the method used in the first generation of DNA sequencers.

Sanger sequencing is the method which prevailed from the 1980s until the mid-2000s. Over that period, great advances were made in the technique, such as fluorescent labelling, capillary electrophoresis, and general automation. These developments allowed much more efficient sequencing, leading to lower costs. The Sanger method, in mass production form, is the technology which produced the first human genome in 2001, ushering in the age of genomics. However, later in the decade, radically different approaches reached the market, bringing the cost per genome down from $100 million in 2001 to $10,000 in 2011.

Advanced Methods and De Novo Sequencing

Large-scale sequencing often aims at sequencing very long DNA pieces, such as whole chromosomes, although large-scale sequencing can also be used to generate very large numbers of short sequences, such as found in phage display. For longer targets such as chromosomes, common approaches consist of cutting (with restriction enzymes) or shearing (with mechanical forces) large DNA fragments into shorter DNA fragments. The fragmented DNA may then be cloned into a DNA vector and amplified in a bacterial host such as *Escherichia coli*. Short DNA fragments purified from individual bacterial colonies are individually sequenced and assembled electronically into one long, contiguous sequence. Studies have shown that adding a size selection step to collect DNA fragments of uniform size can improve sequencing efficiency and accuracy of the genome assembly. In these studies, automated sizing has proven to be more reproducible and precise than manual gel sizing.

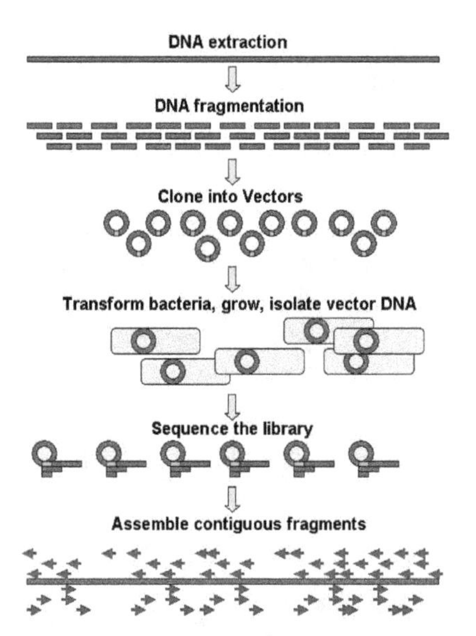

Genomic DNA is fragmented into random pieces and cloned as a bacterial library. DNA from individual bacterial clones is sequenced and the sequence is assembled by using overlapping DNA regions.(click to expand)

The term "*de novo* sequencing" specifically refers to methods used to determine the sequence of DNA with no previously known sequence. *De novo* translates from Latin as "from the beginning". Gaps in the assembled sequence may be filled by primer walking. The different strategies have different tradeoffs in speed and accuracy; shotgun methods are often used for sequencing large genomes, but its assembly is complex and difficult, particularly with sequence repeats often causing gaps in genome assembly.

Most sequencing approaches use an *in vitro* cloning step to amplify individual DNA molecules, because their molecular detection methods are not sensitive enough for single molecule sequencing. Emulsion PCR isolates individual DNA molecules along with primer-coated beads in aqueous droplets within an oil phase. A polymerase chain reaction (PCR) then coats each bead with clonal copies of the DNA molecule followed by immobilization for later sequencing. Emulsion PCR is used in the methods developed by Marguilis et al. (commercialized by 454 Life Sciences), Shendure and Porreca et al. (also known as "Polony sequencing") and SOLiD sequencing, (developed by Agencourt, later Applied Biosystems, now Life Technologies).

Shotgun Sequencing

Shotgun sequencing is a sequencing method designed for analysis of DNA sequences longer than 1000 base pairs, up to and including entire chromosomes. This method requires the target DNA to be broken into random fragments. After sequencing individual fragments, the sequences can be reassembled on the basis of their overlapping regions.

Bridge PCR

Another method for *in vitro* clonal amplification is bridge PCR, in which fragments are amplified upon primers attached to a solid surface and form "DNA colonies" or "DNA clusters". This method is used in the Illumina Genome Analyzer sequencers. Single-molecule methods, such as that developed by Stephen Quake's laboratory (later commercialized by Helicos) are an exception: they use bright fluorophores and laser excitation to detect base addition events from individual DNA molecules fixed to a surface, eliminating the need for molecular amplification.

Next-generation Methods

Next-generation sequencing applies to genome sequencing, genome resequencing, transcriptome profiling (RNA-Seq), DNA-protein interactions (ChIP-sequencing), and epigenome characterization. Resequencing is necessary, because the genome of a single individual of a species will not indicate all of the genome variations among other individuals of the same species.

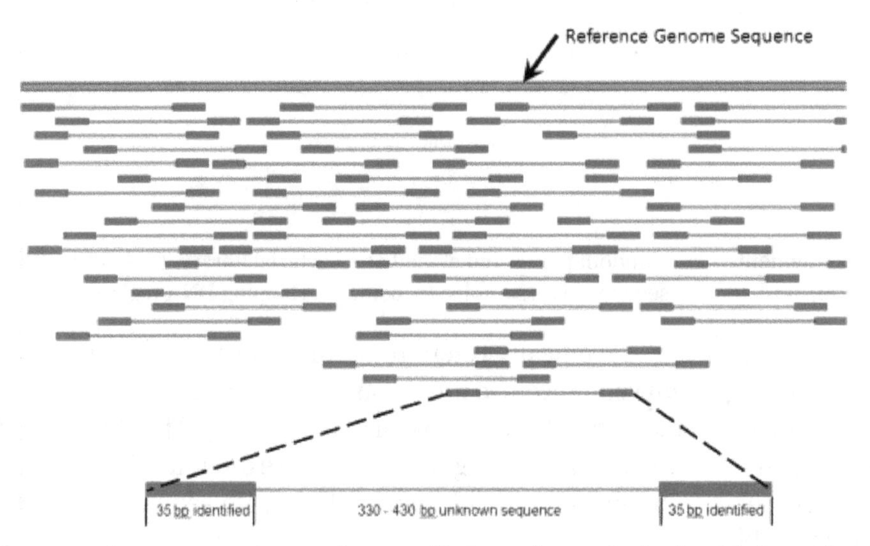

Multiple, fragmented sequence reads must be assembled together on the basis of their overlapping areas.

The high demand for low-cost sequencing has driven the development of high-throughput sequencing (or next-generation sequencing) technologies that parallelize the sequencing process, producing thousands or millions of sequences concurrently. High-throughput sequencing technologies are intended to lower the cost of DNA sequencing beyond what is possible with standard dye-terminator methods. In ultra-high-throughput sequencing as many as 500,000 sequencing-by-synthesis operations may be run in parallel.

Comparison of next-generation sequencing methods							
Method	**Read length**	**Accuracy (single read not consensus)**	**Reads per run**	**Time per run**	**Cost per 1 million bases (in US$)**	**Advantages**	**Disadvantages**
Single-molecule real-time sequencing (Pacific Biosciences)	10,000 bp to 15,000 bp avg (14,000 bp N50); maximum read length >40,000 bases	87% single-read accuracy	50,000 per SMRT cell, or 500–1000 megabases	30 minutes to 4 hours	$0.13–$0.60	Longest read length. Fast. Detects 4mC, 5mC, 6mA.	Moderate throughput. Equipment can be very expensive.

Ion semiconductor (Ion Torrent sequencing)	up to 400 bp	98%	up to 80 million	2 hours	$1	Less expensive equipment. Fast.	Homopolymer errors.
Pyrosequencing (454)	700 bp	99.9%	1 million	24 hours	$10	Long read size. Fast.	Runs are expensive. Homopolymer errors.
Sequencing by synthesis (Illumina)	MiniSeq, NextSeq: 75-300 bp; MiSeq: 50-600 bp; HiSeq 2500: 50-500 bp; HiSeq 3/4000: 50-300 bp; HiSeq X: 300 bp	99.9% (Phred30)	MiniSeq/MiSeq: 1-25 Million; NextSeq: 130-00 Million, HiSeq 2500: 300 million - 2 billion, HiSeq 3/4000 2.5 billion, HiSeq X: 3 billion	1 to 11 days, depending upon sequencer and specified read length	$0.05 to $0.15	Potential for high sequence yield, depending upon sequencer model and desired application.	Equipment can be very expensive. Requires high concentrations of DNA.
Sequencing by ligation (SOLiD sequencing)	50+35 or 50+50 bp	99.9%	1.2 to 1.4 billion	1 to 2 weeks	$0.13	Low cost per base.	Slower than other methods. Has issues sequencing palindromic sequences.
Chain termination (Sanger sequencing)	400 to 900 bp	99.9%	N/A	20 minutes to 3 hours	$2400	Long individual reads. Useful for many applications.	More expensive and impractical for larger sequencing projects. This method also requires the time consuming step of plasmid cloning or PCR.

Massively Parallel Signature Sequencing (MPSS)

The first of the next-generation sequencing technologies, massively parallel signature sequencing (or MPSS), was developed in the 1990s at Lynx Therapeutics, a company founded in 1992 by Sydney Brenner and Sam Eletr. MPSS was a bead-based method that used a complex approach of adapter ligation followed by adapter decoding, reading the sequence in increments of four nucleotides. This method made it susceptible to sequence-specific bias or loss of specific sequences. Because the technology was so complex, MPSS was only performed 'in-house' by Lynx Therapeutics and no DNA sequencing machines were sold to independent laboratories. Lynx Therapeutics merged with Solexa (later acquired by Illumina) in 2004, leading to the development of sequencing-by-synthesis, a simpler approach acquired from Manteia Predictive Medicine, which rendered MPSS obsolete. However, the essential properties of the MPSS output were typical of later "next-generation" data types, including hundreds of thousands of short DNA sequences. In the case of MPSS, these were typically used for sequencing cDNA for measurements of gene expression levels.

Polony Sequencing

The Polony sequencing method, developed in the laboratory of George M. Church at Harvard, was among the first next-generation sequencing systems and was used to sequence a full *E. coli* genome in 2005. It combined an in vitro paired-tag library with emulsion PCR, an automated microscope, and ligation-based sequencing chemistry to sequence an *E. coli* genome at an accuracy of >99.9999% and a cost approximately 1/9 that of Sanger sequencing. The technology was licensed to Agencourt Biosciences, subsequently spun out into Agencourt Personal Genomics, and eventually incorporated into the Applied Biosystems SOLiD platform. Applied Biosystems was later acquired by Life Technologies, now part of Thermo Fisher Scientific.

454 Pyrosequencing

A parallelized version of pyrosequencing was developed by 454 Life Sciences, which has since been acquired by Roche Diagnostics. The method amplifies DNA inside water droplets in an oil solution (emulsion PCR), with each droplet containing a single DNA template attached to a single primer-coated bead that then forms a clonal colony. The sequencing machine contains many picoliter-volume wells each containing a single bead and sequencing enzymes. Pyrosequencing uses luciferase to generate light for detection of the individual nucleotides added to the nascent DNA, and the combined data are used to generate sequence read-outs. This technology provides intermediate read length and price per base compared to Sanger sequencing on one end and Solexa and SOLiD on the other.

Illumina (Solexa) Sequencing

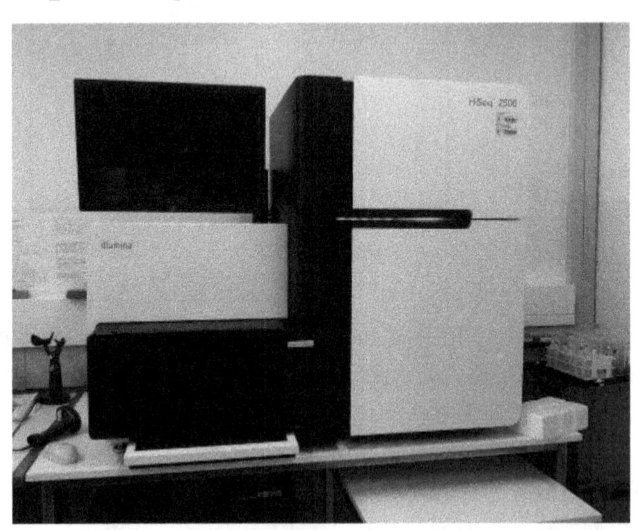

An Illumina HiSeq 2500 sequencer

Solexa, now part of Illumina, was founded by Shankar Balasubramanian and David Klenerman in 1998, and developed a sequencing method based on reversible dye-terminators technology, and engineered polymerases. The reversible terminated chemistry concept was invented by Bruno Canard and Simon Sarfati at the Pasteur Institute in Paris. It was developed internally at Solexa by those named on the relevant patents. In 2004, Solexa acquired the company Manteia Predictive Medicine in order to gain a massivelly parallel sequencing technology invented in 1997 by Pascal Mayer and Laurent Farinelli. It is based on "DNA Clusters" or "DNA colonies", which involves the clonal amplification of DNA on a surface. The cluster technology was co-acquired with Lynx Therapeutics of California. Solexa Ltd. later merged with Lynx to form Solexa Inc.

In this method, DNA molecules and primers are first attached on a slide or flow cell and amplified with polymerase so that local clonal DNA colonies, later coined "DNA clusters", are formed. To determine the sequence, four types of reversible terminator bases (RT-bases) are added and non-incorporated nucleotides are washed away. A camera takes images of the fluorescently labeled nucleotides. Then the dye, along with the terminal 3' blocker, is chemically removed from the DNA, allowing for the next cycle to begin. Unlike pyrosequencing, the DNA chains are extended one nucleotide at a time and image acquisition can be performed at a delayed moment, allowing for very large arrays of DNA colonies to be captured by sequential images taken from a single camera.

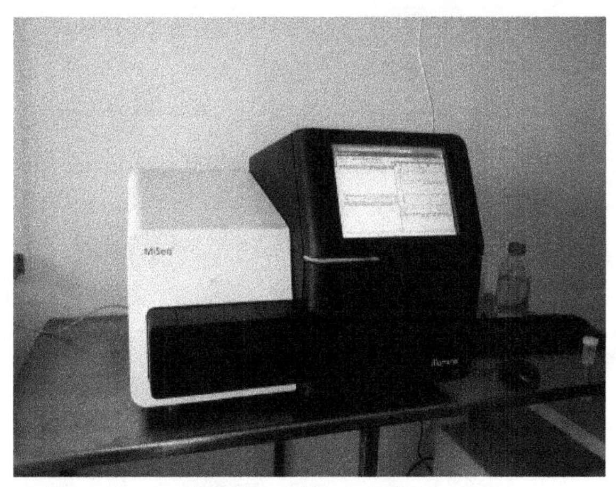

An Illumina MiSeq sequencer

Decoupling the enzymatic reaction and the image capture allows for optimal throughput and theoretically unlimited sequencing capacity. With an optimal configuration, the ultimately reachable instrument throughput is thus dictated solely by the analog-to-digital conversion rate of the camera, multiplied by the number of cameras and divided by the number of pixels per DNA colony required for visualizing them optimally (approximately 10 pixels/colony). In 2012, with cameras operating at more than 10 MHz A/D conversion rates and available optics, fluidics and enzymatics, throughput can be multiples of 1 million nucleotides/second, corresponding roughly to 1 human genome equivalent at 1x coverage per hour per instrument, and 1 human genome re-sequenced (at approx. 30x) per day per instrument (equipped with a single camera).

SOLiD Sequencing

Applied Biosystems' (now a Life Technologies brand) SOLiD technology employs sequencing by ligation. Here, a pool of all possible oligonucleotides of a fixed length are labeled according to the sequenced position. Oligonucleotides are annealed and ligated; the preferential ligation by DNA ligase for matching sequences results in a signal informative of the nucleotide at that position. Before sequencing, the DNA is amplified by emulsion PCR. The resulting beads, each containing single copies of the same DNA molecule, are deposited on a glass slide. The result is sequences of quantities and lengths comparable to Illumina sequencing. This sequencing by ligation method has been reported to have some issue sequencing palindromic sequences.

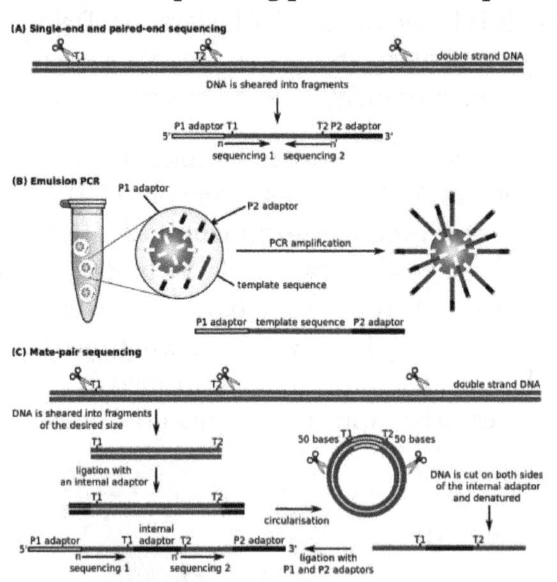

Library preparation for the SOLiD platform

Ion Torrent Semiconductor Sequencing

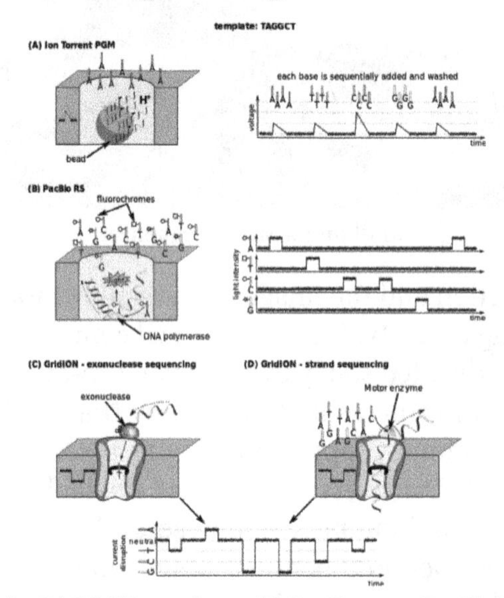

Sequencing of the TAGGCT template with IonTorrent, PacBioRS and GridION

Ion Torrent Systems Inc. (now owned by Life Technologies) developed a system based on using standard sequencing chemistry, but with a novel, semiconductor based detection system. This method of sequencing is based on the detection of hydrogen ions that are released during the polymerisation of DNA, as opposed to the optical methods used in other sequencing systems. A microwell containing a template DNA strand to be sequenced is flooded with a single type of nucleotide. If the introduced nucleotide is complementary to the leading template nucleotide it is incorporated into the growing complementary strand. This causes the release of a hydrogen ion that triggers a hypersensitive ion sensor, which indicates that a reaction has occurred. If homopolymer repeats are present in the template sequence multiple nucleotides will be incorporated in a single cycle. This leads to a corresponding number of released hydrogens and a proportionally higher electronic signal.

DNA Nanoball Sequencing

DNA nanoball sequencing is a type of high throughput sequencing technology used to determine the entire genomic sequence of an organism. The company Complete Genomics uses this technology to sequence samples submitted by independent researchers. The method uses rolling circle replication to amplify small fragments of genomic DNA into DNA nanoballs. Unchained sequencing by ligation is then used to determine the nucleotide sequence. This method of DNA sequencing allows large numbers of DNA nanoballs to be sequenced per run and at low reagent costs compared to other next generation sequencing platforms. However, only short sequences of DNA are determined from each DNA nanoball which makes mapping the short reads to a reference genome difficult. This technology has been used for multiple genome sequencing projects and is scheduled to be used for more.

Heliscope Single Molecule Sequencing

Heliscope sequencing is a method of single-molecule sequencing developed by Helicos Biosciences. It uses DNA fragments with added poly-A tail adapters which are attached to the flow cell surface. The next steps involve extension-based sequencing with cyclic washes of the flow cell with fluorescently labeled nucleotides (one nucleotide type at a time, as with the Sanger method). The reads are performed by the Heliscope sequencer. The reads are short, averaging 35 bp. In 2009 a human genome was sequenced using the Heliscope, however in 2012 the company went bankrupt.

Single Molecule Real Time (SMRT) Sequencing

SMRT sequencing is based on the sequencing by synthesis approach. The DNA is synthesized in zero-mode wave-guides (ZMWs) – small well-like containers with the capturing tools located at the bottom of the well. The sequencing is performed with use of unmodified polymerase (attached to the ZMW bottom) and fluorescently labelled nucleotides flowing freely in the solution. The wells are constructed in a way that only the fluorescence occurring by the bottom of the well is detected. The fluorescent label is detached from the nucleotide upon its incorporation into the DNA strand, leaving an unmodified DNA strand. According to Pacific Biosciences (PacBio), the SMRT technology developer, this methodology allows detection of nucleotide modifications (such as cytosine methylation). This happens through the observation of polymerase kinetics. This approach allows reads of 20,000 nucleotides or more, with average read lengths of 5 kilobases. In 2015, Pacific Biosciences announced the launch of a new sequencing instrument called the Sequel System, with 1 million ZMWs compared to 150,000 ZMWs in the PacBio RS II instrument.

Nanopore DNA Sequencing

The DNA passing through the nanopore changes its ion current. This change is dependent on the shape, size and length of the DNA sequence. Each type of the nucleotide blocks the ion flow through the pore for a different period of time. The method does not require modified nucleotides and is performed in real time.

Early industrial research into this method was based on a technique called 'Exonuclease sequencing', where the readout of electrical signals occurring at nucleotides passing by alpha-hemolysin pores covalently bound with cyclodextrin. However the subsequently commercial method, 'strand sequencing' sequencing DNA bases in an intact strand. In 2014, Oxford Nanopore Technologies, a United Kingdom-based company, released the MinION portable DNA sequencer into an early access community and the first dataset became available in June 2014. The MinION was subsequently launched commercially in 2015 and the launch of the 'R9' nanopore in 2016 resulted in reports of performance improvements in accuracy and yield. in autumn 2016, the MinION handheld sequencer is capable of generating more than 2 Gigabases of sequencing data in one run. it is used in multiple scientific applications such as pathogen analysis/surveillance, metagenomics, cancer research and environmental analysis and a version has also been recently installed on the ISS by NASA.

Two main areas of nanopore sequencing in development are solid state nanopore sequencing, and protein based nanopore sequencing. Protein nanopore sequencing utilizes membrane protein complexes such as ∝-Hemolysin, MspA (Mycobacterium Smegmatis Porin A) or CssG, which show great promise given their ability to distinguish between individual and groups of nucleotides. In contrast, solid-state nanopore sequencing utilizes synthetic materials such as silicon nitride and aluminum oxide and it is preferred for its superior mechanical ability and thermal and chemical stability. The fabrication method is essential for this type of sequencing given that the nanopore array can contain hundreds of pores with diameters smaller than eight nanometers.

The concept originated from the idea that single stranded DNA or RNA molecules can be electrophoretically driven in a strict linear sequence through a biological pore that can be less than eight nanometers, and can be detected given that the molecules release an ionic current while moving through the pore. The pore contains a detection region capable of recognizing different bases, with each base generating various time specific signals corresponding to the sequence of bases as they cross the pore which are then evaluated. Precise control over the DNA transport through the pore is crucial for success. Various enzymes such as exonucleases and polymerases have been used to moderate this process by positioning them near the pore's entrance.

Methods in Development

DNA sequencing methods currently under development include reading the sequence as a DNA strand transits through nanopores (a method that is now commercial but subsequent generations such as solid-state nanopores are still in development), and microscopy-based techniques, such as atomic force microscopy or transmission electron microscopy that are used to identify the positions of individual nucleotides within long DNA fragments (>5,000 bp) by nucleotide labeling with heavier elements (e.g., halogens) for visual detection and recording. Third generation technologies aim to increase throughput and decrease the time to result and cost by eliminating the need for excessive reagents and harnessing the processivity of DNA polymerase.

Tunnelling Currents DNA Sequencing

Another approach uses measurements of the electrical tunnelling currents across single-strand

DNA as it moves through a channel. Depending on its electronic structure, each base affects the tunnelling current differently, allowing differentiation between different bases.

The use of tunnelling currents has the potential to sequence orders of magnitude faster than ionic current methods and the sequencing of several DNA oligomers and micro-RNA has already been achieved.

Sequencing by Hybridization

Sequencing by hybridization is a non-enzymatic method that uses a DNA microarray. A single pool of DNA whose sequence is to be determined is fluorescently labeled and hybridized to an array containing known sequences. Strong hybridization signals from a given spot on the array identifies its sequence in the DNA being sequenced.

This method of sequencing utilizes binding characteristics of a library of short single stranded DNA molecules (oligonucleotides), also called DNA probes, to reconstruct a target DNA sequence. Non-specific hybrids are removed by washing and the target DNA is eluted. Hybrids are re-arranged such that the DNA sequence can be reconstructed. The benefit of this sequencing type is its ability to capture a large number of targets with a homogenous coverage. A large number of chemicals and starting DNA is usually required. However, with the advent of solution-based hybridization, much less equipment and chemicals are necessary.

Sequencing with Mass Spectrometry

Mass spectrometry may be used to determine DNA sequences. Matrix-assisted laser desorption ionization time-of-flight mass spectrometry, or MALDI-TOF MS, has specifically been investigated as an alternative method to gel electrophoresis for visualizing DNA fragments. With this method, DNA fragments generated by chain-termination sequencing reactions are compared by mass rather than by size. The mass of each nucleotide is different from the others and this difference is detectable by mass spectrometry. Single-nucleotide mutations in a fragment can be more easily detected with MS than by gel electrophoresis alone. MALDI-TOF MS can more easily detect differences between RNA fragments, so researchers may indirectly sequence DNA with MS-based methods by converting it to RNA first.

The higher resolution of DNA fragments permitted by MS-based methods is of special interest to researchers in forensic science, as they may wish to find single-nucleotide polymorphisms in human DNA samples to identify individuals. These samples may be highly degraded so forensic researchers often prefer mitochondrial DNA for its higher stability and applications for lineage studies. MS-based sequencing methods have been used to compare the sequences of human mitochondrial DNA from samples in a Federal Bureau of Investigation database and from bones found in mass graves of World War I soldiers.

Early chain-termination and TOF MS methods demonstrated read lengths of up to 100 base pairs. Researchers have been unable to exceed this average read size; like chain-termination sequencing alone, MS-based DNA sequencing may not be suitable for large *de novo* sequencing projects. Even so, a recent study did use the short sequence reads and mass spectroscopy to compare single-nucleotide polymorphisms in pathogenic *Streptococcus* strains.

Microfluidic Sanger Sequencing

In microfluidic Sanger sequencing the entire thermocycling amplification of DNA fragments as well as their separation by electrophoresis is done on a single glass wafer (approximately 10 cm in diameter) thus reducing the reagent usage as well as cost. In some instances researchers have shown that they can increase the throughput of conventional sequencing through the use of microchips. Research will still need to be done in order to make this use of technology effective.

Microscopy-based Techniques

This approach directly visualizes the sequence of DNA molecules using electron microscopy. The first identification of DNA base pairs within intact DNA molecules by enzymatically incorporating modified bases, which contain atoms of increased atomic number, direct visualization and identification of individually labeled bases within a synthetic 3,272 base-pair DNA molecule and a 7,249 base-pair viral genome has been demonstrated.

RNAP Sequencing

This method is based on use of RNA polymerase (RNAP), which is attached to a polystyrene bead. One end of DNA to be sequenced is attached to another bead, with both beads being placed in optical traps. RNAP motion during transcription brings the beads in closer and their relative distance changes, which can then be recorded at a single nucleotide resolution. The sequence is deduced based on the four readouts with lowered concentrations of each of the four nucleotide types, similarly to the Sanger method. A comparison is made between regions and sequence information is deduced by comparing the known sequence regions to the unknown sequence regions.

In Vitro Virus High-throughput Sequencing

A method has been developed to analyze full sets of protein interactions using a combination of 454 pyrosequencing and an *in vitro* virus mRNA display method. Specifically, this method covalently links proteins of interest to the mRNAs encoding them, then detects the mRNA pieces using reverse transcription PCRs. The mRNA may then be amplified and sequenced. The combined method was titled IVV-HiTSeq and can be performed under cell-free conditions, though its results may not be representative of *in vivo* conditions.

Sample Preparation

The success of a DNA sequencing protocol is dependent on the sample preparation. A successful DNA extraction will yield a sample with long, non-degraded strands of DNA which require further preparation according to the sequencing technology to be used. For Sanger sequencing, either cloning procedures or PCR are required prior to sequencing. In the case of next generation sequencing methods, library preparation is required before processing.

With the advent of next generation sequencing, Illumina and Roche 454 methods have become a common approach to transcriptomic studies (RNAseq). RNA can be extracted from tissues of interest and converted to complimentary DNA (cDNA) using reverse transcriptase—a DNA polymerase that synthesizes a complimentary DNA based on existing strands of RNA in a PCR-like manner. Complimentary DNA can be processed the same way as genomic DNA, allowing the expression levels of RNAs to be determined for the tissue selected.

Development Initiatives

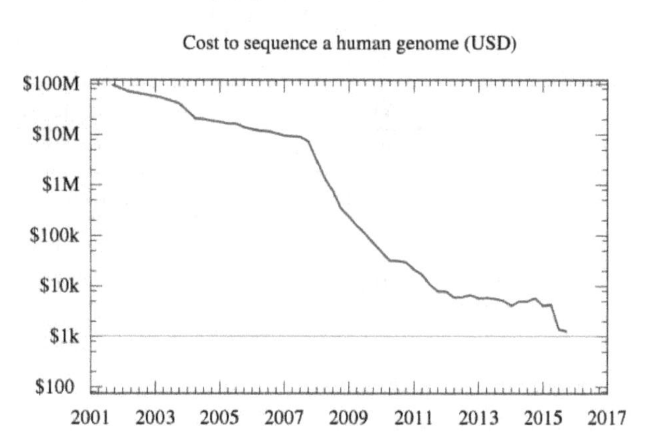

Total cost of sequencing a human genome over time as calculated by the NHGRI.

In October 2006, the X Prize Foundation established an initiative to promote the development of full genome sequencing technologies, called the Archon X Prize, intending to award $10 million to "the first Team that can build a device and use it to sequence 100 human genomes within 10 days or less, with an accuracy of no more than one error in every 100,000 bases sequenced, with sequences accurately covering at least 98% of the genome, and at a recurring cost of no more than $10,000 (US) per genome."

Each year the National Human Genome Research Institute, or NHGRI, promotes grants for new research and developments in genomics. 2010 grants and 2011 candidates include continuing work in microfluidic, polony and base-heavy sequencing methodologies.

Computational Challenges

The sequencing technologies described here produce raw data that needs to be assembled into longer sequences such as complete genomes (sequence assembly). There are many computational challenges to achieve this, such as the evaluation of the raw sequence data which is done by programs and algorithms such as Phred and Phrap. Other challenges have to deal with repetitive sequences that often prevent complete genome assemblies because they occur in many places of the genome. As a consequence, many sequences may not be assigned to particular chromosomes. The production of raw sequence data is only the beginning of its detailed bioinformatical analysis. Yet new methods for sequencing and correcting sequencing errors were developed.

Read Trimming

Sometimes, the raw reads produced by the sequencer are correct and precise only in a fraction of their length. Using the entire read may introduce artifacts in the downstream analyses like genome assembly, snp calling, or gene expression estimation. Two classes of trimming programs have been introduced, based on the window-based or the running-sum classes of algorithms. This is a partial list of the trimming algorithms currently available, specifying the algorithm class they belong to:

Read Trimming Algorithms		
Name of algorithm	**Type of algorithm**	**Link**
Cutadapt	Running sum	Cutadapt
ConDeTri	Window based	ConDeTri
ERNE-FILTER	Running sum	ERNE-FILTER
FASTX quality trimmer	Window based	FASTX quality trimmer
PRINSEQ	Window based	PRINSEQ
Trimmomatic	Window based	Trimmomatic
SolexaQA	Window based	SolexaQA
SolexaQA-BWA	Running sum	SolexaQA-BWA
Sickle	Window based	Sickle

Ethical Issues

Human genetics have been included within the field of bioethics since the early 1970s and the growth in the use of DNA sequencing (particularly high-throughput sequencing) has introduced a number of ethical issues. One key issue is the ownership of an individual's DNA and the data produced when that DNA is sequenced. Regarding the DNA molecule itself, the leading legal case on this topic, *Moore v. Regents of the University of California* (1990) ruled that individuals have no property rights to discarded cells or any profits made using these cells (for instance, as a patented cell line). However, individuals have a right to informed consent regarding removal and use of cells. Regarding the data produced through DNA sequencing, *Moore* gives the individual no rights to the information derived from their DNA.

As DNA sequencing becomes more widespread, the storage, security and sharing of genomic data has also become more important. For instance, one concern is that insurers may use an individual's genomic data to modify their quote, depending on the perceived future health of the individual based on their DNA. In May 2008, the Genetic Information Nondiscrimination Act (GINA) was signed in the United States, prohibiting discrimination on the basis of genetic information with respect to health insurance and employment. In 2012, the US Presidential Commission for the Study of Bioethical Issues reported that existing privacy legislation for DNA sequencing data such as GINA and the Health Insurance Portability and Accountability Act were insufficient, noting that whole-genome sequencing data was particularly sensitive, as it could be used to identify not only the individual from which the data was created, but also their relatives.

Ethical issues have also been raised by the increasing use of genetic variation screening, both in newborns, and in adults by companies such as 23andMe. It has been asserted that screening for genetic variations can be harmful, increasing anxiety in individuals who have been found to have an increased risk of disease. For example, in one case noted in *Time*, doctors screening an ill baby for genetic variants chose not to inform the parents of an unrelated variant linked to dementia due to the harm it would cause to the parents. However, a 2011 study in *The New England Journal of Medicine* has shown that individuals undergoing disease risk profiling did not show increased levels of anxiety.

DNA Microarray

A DNA microarray (also commonly known as DNA chip or biochip) is a collection of microscopic DNA spots attached to a solid surface. Scientists use DNA microarrays to measure the expression levels of large numbers of genes simultaneously or to genotype multiple regions of a genome. Each DNA spot contains picomoles (10^{-12} moles) of a specific DNA sequence, known as *probes* (or *reporters* or *oligos*). These can be a short section of a gene or other DNA element that are used to hybridize a cDNA or cRNA (also called anti-sense RNA) sample (called *target*) under high-stringency conditions. Probe-target hybridization is usually detected and quantified by detection of fluorophore-, silver-, or chemiluminescence-labeled targets to determine relative abundance of nucleic acid sequences in the target.

Principle

The core principle behind microarrays is hybridization between two DNA strands, the property of complementary nucleic acid sequences to specifically pair with each other by forming hydrogen bonds between complementary nucleotide base pairs. A high number of complementary base pairs in a nucleotide sequence means tighter non-covalent bonding between the two strands. After washing off non-specific bonding sequences, only strongly paired strands will remain hybridized. Fluorescently labeled target sequences that bind to a probe sequence generate a signal that depends on the hybridization conditions (such as temperature), and washing after hybridization. Total strength of the signal, from a spot (feature), depends upon the amount of target sample binding to the probes present on that spot. Microarrays use relative quantitation in which the intensity of a feature is compared to the intensity of the same feature under a different condition, and the identity of the feature is known by its position.

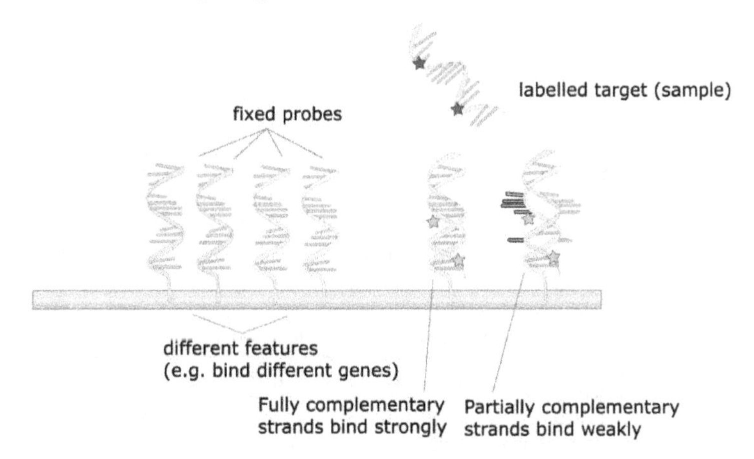

Hybridization of the target to the probe.

Uses and Types

Many types of arrays exist and the broadest distinction is whether they are spatially arranged on a surface or on coded beads:

- The traditional solid-phase array is a collection of orderly microscopic "spots", called features, each with thousands of identical and specific probes attached to a solid surface, such as glass, plastic or silicon biochip (commonly known as a *genome chip, DNA chip* or *gene array*). Thousands of these features can be placed in known locations on a single DNA microarray.

- The alternative bead array is a collection of microscopic polystyrene beads, each with a specific probe and a ratio of two or more dyes, which do not interfere with the fluorescent dyes used on the target sequence.

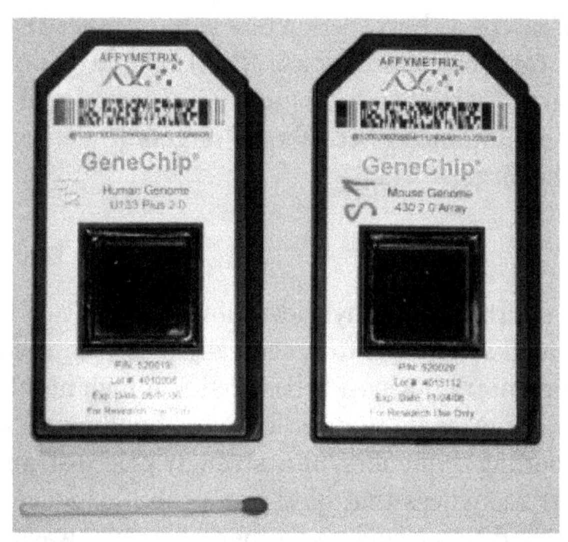

Two Affymetrix chips. A match is shown at bottom left for size comparison.

DNA microarrays can be used to detect DNA (as in comparative genomic hybridization), or detect RNA (most commonly as cDNA after reverse transcription) that may or may not be translated into proteins. The process of measuring gene expression via cDNA is called expression analysis or expression profiling.

Applications include:

Application or technology	Synopsis
Gene expression profiling	In an mRNA or gene expression profiling experiment the expression levels of thousands of genes are simultaneously monitored to study the effects of certain treatments, diseases, and developmental stages on gene expression. For example, microarray-based gene expression profiling can be used to identify genes whose expression is changed in response to pathogens or other organisms by comparing gene expression in infected to that in uninfected cells or tissues.
Comparative genomic hybridization	Assessing genome content in different cells or closely related organisms.
GeneID	Small microarrays to check IDs of organisms in food and feed (like GMO), mycoplasms in cell culture, or pathogens for disease detection, mostly combining PCR and microarray technology.
Chromatin immunoprecipitation on Chip	DNA sequences bound to a particular protein can be isolated by immunoprecipitating that protein (ChIP), these fragments can be then hybridized to a microarray (such as a tiling array) allowing the determination of protein binding site occupancy throughout the genome. Example protein to immunoprecipitate are histone modifications (H3K27me3, H3K4me2, H3K9me3, etc.), Polycomb-group protein (PRC2:Suz12, PRC1:YY1) and trithorax-group protein (Ash1) to study the epigenetic landscape or RNA Polymerase II to study the transcription landscape.

DamID	Analogously to ChIP, genomic regions bound by a protein of interest can be isolated and used to probe a microarray to determine binding site occupancy. Unlike ChIP, DamID does not require antibodies but makes use of adenine methylation near the protein's binding sites to selectively amplify those regions, introduced by expressing minute amounts of protein of interest fused to bacterial DNA adenine methyltransferase.
SNP detection	Identifying single nucleotide polymorphism among alleles within or between populations. Several applications of microarrays make use of SNP detection, including genotyping, forensic analysis, measuring predisposition to disease, identifying drug-candidates, evaluating germline mutations in individuals or somatic mutations in cancers, assessing loss of heterozygosity, or genetic linkage analysis.
Alternative splicing detection	An *exon junction array* design uses probes specific to the expected or potential splice sites of predicted exons for a gene. It is of intermediate density, or coverage, to a typical gene expression array (with 1-3 probes per gene) and a genomic tiling array (with hundreds or thousands of probes per gene). It is used to assay the expression of alternative splice forms of a gene. Exon arrays have a different design, employing probes designed to detect each individual exon for known or predicted genes, and can be used for detecting different splicing isoforms.
Fusion genes microarray	A Fusion gene microarray can detect fusion transcripts, *e.g.* from cancer specimens. The principle behind this is building on the alternative splicing microarrays. The oligo design strategy enables combined measurements of chimeric transcript junctions with exon-wise measurements of individual fusion partners.
Tiling array	Genome tiling arrays consist of overlapping probes designed to densely represent a genomic region of interest, sometimes as large as an entire human chromosome. The purpose is to empirically detect expression of transcripts or alternatively spliced forms which may not have been previously known or predicted.
Double-stranded B-DNA microarrays	Right-handed double-stranded B-DNA microarrays can be used to characterize novel drugs and biologicals that can be employed to bind specific regions of immobilized, intact, double-stranded DNA. This approach can be used to inhibit gene expression. They also allow for characterization of their structure under different environmental conditions.
Double-stranded Z-DNA microarrays	Left-handed double-stranded Z-DNA microarrays can be used to identify short sequences of the alternative Z-DNA structure located within longer stretches of right-handed B-DNA genes (e.g., transcriptional enhancement, recombination, RNA editing). The microarrays also allow for characterization of their structure under different environmental conditions.
Multi-stranded DNA microarrays (triplex-DNA microarrays and quadruplex-DNA microarrays)	Multi-stranded DNA and RNA microarrays can be used to identify novel drugs that bind to these multi-stranded nucleic acid sequences. This approach can be used to discover new drugs and biologicals that have the ability to inhibit gene expression. These microarrays also allow for characterization of their structure under different environmental conditions.

Fabrication

Microarrays can be manufactured in different ways, depending on the number of probes under examination, costs, customization requirements, and the type of scientific question being asked. Arrays may have as few as 10 probes or up to 2.1 million micrometre-scale probes from commercial vendors.

Spotted Vs. in Situ Synthesised Arrays

Play mediaA DNA microarray being printed by a robot at the University of Delaware

Microarrays can be fabricated using a variety of technologies, including printing with fine-pointed pins onto glass slides, photolithography using pre-made masks, photolithography using dynamic micromirror devices, ink-jet printing, or electrochemistry on microelectrode arrays.

In *spotted microarrays*, the probes are oligonucleotides, cDNA or small fragments of PCR products that correspond to mRNAs. The probes are synthesized prior to deposition on the array surface and are then "spotted" onto glass. A common approach utilizes an array of fine pins or needles controlled by a robotic arm that is dipped into wells containing DNA probes and then depositing each probe at designated locations on the array surface. The resulting "grid" of probes represents the nucleic acid profiles of the prepared probes and is ready to receive complementary cDNA or cRNA "targets" derived from experimental or clinical samples. This technique is used by research scientists around the world to produce "in-house" printed microarrays from their own labs. These arrays may be easily customized for each experiment, because researchers can choose the probes and printing locations on the arrays, synthesize the probes in their own lab (or collaborating facility), and spot the arrays. They can then generate their own labeled samples for hybridization, hybridize the samples to the array, and finally scan the arrays with their own equipment. This provides a relatively low-cost microarray that may be customized for each study, and avoids the costs of purchasing often more expensive commercial arrays that may represent vast numbers of genes that are not of interest to the investigator. Publications exist which indicate in-house spotted microarrays may not provide the same level of sensitivity compared to commercial oligonucleotide arrays, possibly owing to the small batch sizes and reduced printing efficiencies when compared to industrial manufactures of oligo arrays.

In *oligonucleotide microarrays*, the probes are short sequences designed to match parts of the sequence of known or predicted open reading frames. Although oligonucleotide probes are often used in "spotted" microarrays, the term "oligonucleotide array" most often refers to a specific technique of manufacturing. Oligonucleotide arrays are produced by printing short oligonucleotide sequences designed to represent a single gene or family of gene splice-variants by synthesizing this sequence directly onto the array surface instead of depositing intact sequences. Sequences may be longer (60-

mer probes such as the Agilent design) or shorter (25-mer probes produced by Affymetrix) depending on the desired purpose; longer probes are more specific to individual target genes, shorter probes may be spotted in higher density across the array and are cheaper to manufacture. One technique used to produce oligonucleotide arrays include photolithographic synthesis (Affymetrix) on a silica substrate where light and light-sensitive masking agents are used to "build" a sequence one nucleotide at a time across the entire array. Each applicable probe is selectively "unmasked" prior to bathing the array in a solution of a single nucleotide, then a masking reaction takes place and the next set of probes are unmasked in preparation for a different nucleotide exposure. After many repetitions, the sequences of every probe become fully constructed. More recently, Maskless Array Synthesis from NimbleGen Systems has combined flexibility with large numbers of probes.

Two-channel Vs. One-channel Detection

Two-color microarrays or *two-channel microarrays* are typically hybridized with cDNA prepared from two samples to be compared (e.g. diseased tissue versus healthy tissue) and that are labeled with two different fluorophores. Fluorescent dyes commonly used for cDNA labeling include Cy3, which has a fluorescence emission wavelength of 570 nm (corresponding to the orange part of the light spectrum), and Cy5 with a fluorescence emission wavelength of 670 nm (corresponding to the red part of the light spectrum). The two Cy-labeled cDNA samples are mixed and hybridized to a single microarray that is then scanned in a microarray scanner to visualize fluorescence of the two fluorophores after excitation with a laser beam of a defined wavelength. Relative intensities of each fluorophore may then be used in ratio-based analysis to identify up-regulated and down-regulated genes.

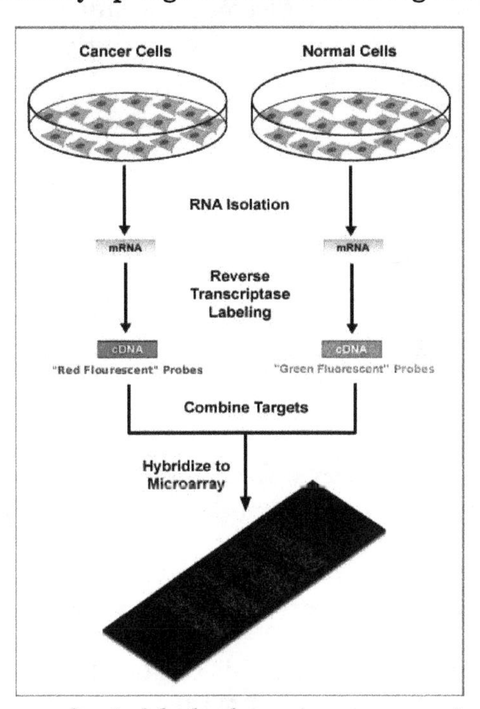

Diagram of typical dual-colour microarray experiment.

Oligonucleotide microarrays often carry control probes designed to hybridize with RNA spike-ins. The degree of hybridization between the spike-ins and the control probes is used to normalize the hybridization measurements for the target probes. Although absolute levels of gene expression may be determined in the two-color array in rare instances, the relative differences in expression among different spots within a sample and between samples is the preferred method of data analysis for the two-color system. Examples of providers for such microarrays includes Agilent with their Dual-Mode platform, Eppendorf with their DualChip platform for colorimetric Silverquant labeling, and TeleChem International with Arrayit.

In *single-channel microarrays* or *one-color microarrays*, the arrays provide intensity data for each probe or probe set indicating a relative level of hybridization with the labeled target. However, they do not truly indicate abundance levels of a gene but rather relative abundance when compared to other samples or conditions when processed in the same experiment. Each RNA molecule encounters protocol and batch-specific bias during amplification, labeling, and hybridization phases of the experiment making comparisons between genes for the same microarray uninformative. The comparison of two conditions for the same gene requires two separate single-dye hybridizations. Several popular single-channel systems are the Affymetrix "Gene Chip", Illumina "Bead Chip", Agilent single-channel arrays, the Applied Microarrays "CodeLink" arrays, and the Eppendorf "DualChip & Silverquant". One strength of the single-dye system lies in the fact that an aberrant sample cannot affect the raw data derived from other samples, because each array chip is exposed to only one sample (as opposed to a two-color system in which a single low-quality sample may drastically impinge on overall data precision even if the other sample was of high quality). Another benefit is that data are more easily compared to arrays from different experiments as long as batch effects have been accounted for.

One channel microarray may be the only choice in some situations. Suppose samples need to be compared: then the number of experiments required using the two channel arrays quickly becomes unfeasible, unless a sample is used as a reference.

number of samples	one-channel microarray	two channel microarray	two channel microarray (with reference)
1	1	1	1
2	2	1	1
3	3	3	2
4	4	6	3

A Typical Protocol

This is an example of a **DNA microarray experiment**, detailing a particular case to better explain DNA microarray experiments, while enumerating possible alternatives.

1. The two samples to be compared (pairwise comparison) are grown/acquired. In this example treated sample (case) and untreated sample (control).

2. The nucleic acid of interest is purified: this can be all RNA for expression profiling, DNA for comparative hybridization, or DNA/RNA bound to a particular protein which is immunoprecipitated (ChIP-on-chip) for epigenetic or regulation studies. In this example total RNA is isolated (total as it is nuclear and cytoplasmic) by Guanidinium thiocyanate-phenol-chloroform extraction (e.g. Trizol) which isolates most RNA (whereas column methods have a cut off of 200 nucleotides) and if done correctly has a better purity.

3. The purified RNA is analysed for quality (by capillary electrophoresis) and quantity (for example, by using a NanoDrop or NanoPhotometer spectrometer). If the material is of acceptable quality and sufficient quantity is present (e.g., >1μg, although the required amount varies by microarray platform), the experiment can proceed.

4. The labelled product is generated via reverse transcription and sometimes with an optional PCR amplification. The RNA is reverse transcribed with either polyT primers (which amplify only mRNA) or random primers (which amplify all RNA, most of which is rRNA); miRNA microarrays ligate an oligonucleotide to the purified small RNA (isolated with a fractionator), which is then reverse transcribed and amplified. The label is added either during the reverse transcription step, or following amplification if it is performed. The sense labelling is dependent on the microarray; e.g. if the label is added with the RT mix, the cDNA is antisense and the microarray probe is sense, except in the case of negative controls. The label is typically fluorescent; only one machine uses radiolabels. The labelling can be direct (not used) or indirect (requires a coupling stage). For two-channel arrays, the coupling stage occurs before hybridization, using aminoallyl uridine triphosphate (aminoallyl-UTP, or aaUTP) and NHS amino-reactive dyes (such as cyanine dyes); for single-channel arrays, the coupling stage occurs after hybridization, using biotin and labelled streptavin. The modified nucleotides (usually in a ratio of 1 aaUTP: 4 TTP (thymidine triphosphate)) are added enzymatically in a low ratio to normal nucleotides, typically resulting in 1 every 60 bases. The aaDNA is then purified with a column (using a phosphate buffer solution, as Tris contains amine groups). The aminoallyl group is an amine group on a long linker attached to the nucleobase, which reacts with a reactive dye. A form of replicate known as a dye flipcan be performed to remove any dye effects in two-channel experiments; for a dye flip, a second slide is used, with the labels swapped (the sample that was labeled with Cy3 in the first slide is labeled with Cy5, and vice versa). In this example, aminoallyl-UTP is present in the reverse-transcribed mixture.

5. The labeled samples are then mixed with a propriety hybridization solution which can consist of SDS, SSC, dextran sulfate, a blocking agent (such as COT1 DNA, salmon sperm DNA, calf thymus DNA, PolyA or PolyT), Denhardt's solution, or formamine.

6. The mixture is denatured and added to the pinholes of the microarray. The holes are sealed and the microarray hybridized, either in a hyb oven, where the microarray is mixed by rotation, or in a mixer, where the microarray is mixed by alternating pressure at the pinholes.

7. After an overnight hybridization, all nonspecific binding is washed off (SDS and SSC).

8. The microarray is dried and scanned by a machine that uses a laser to excite the dye and measures the emission levels with a detector.

9. The image is gridded with a template and the intensities of each feature (composed of several pixels) is quantified.

10. The raw data is normalized; the simplest normalization method is to subtract background intensity and scale so that the total intensities of the features of the two channels are equal, or to use the intensity of a reference gene to calculate the t-value for all of the intensities. More sophisticated methods include z-ratio, loess and lowess regression and RMA (robust

multichip analysis) for Affymetrix chips (single-channel, silicon chip, in situ synthesised short oligonucleotides).

Microarrays and Bioinformatics

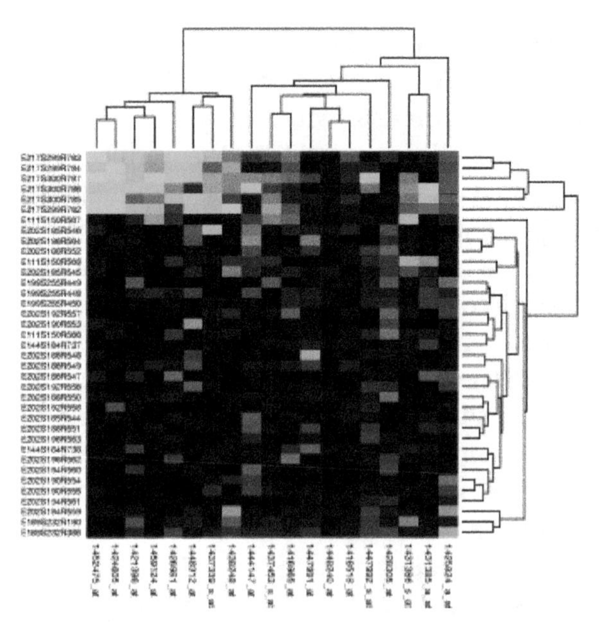

Gene expression values from microarray experiments can be represented as heat maps to visualize the result of data analysis.

The advent of inexpensive microarray experiments created several specific bioinformatics challenges:

- the multiple levels of replication in experimental design (Experimental design)
- the number of platforms and independent groups and data format (Standardization)
- the statistical treatment of the data (Statistical analysis)
- mapping each probe to the mRNA transcript that it measures (Annotation)
- the sheer volume of data and the ability to share it (Data warehousing)

Experimental Design

Due to the biological complexity of gene expression, the considerations of experimental design that are discussed in the expression profiling article are of critical importance if statistically and biologically valid conclusions are to be drawn from the data.

There are three main elements to consider when designing a microarray experiment. First, replication of the biological samples is essential for drawing conclusions from the experiment. Second, technical replicates (two RNA samples obtained from each experimental unit) help to ensure precision and allow for testing differences within treatment groups. The biological replicates include independent RNA extractions and technical replicates may be two aliquots of the same extraction. Third, spots of each cDNA clone or oligonucleotide are present as replicates (at least duplicates) on the microarray slide, to provide a measure of technical precision in each hybridization. It is critical that information about the sample preparation and handling is discussed, in order to help identify the independent units in the experiment and to avoid inflated estimates of statistical significance.

Standardization

Microarray data is difficult to exchange due to the lack of standardization in platform fabrication, assay protocols, and analysis methods. This presents an interoperability problem in bioinformatics. Various grass-roots open-source projects are trying to ease the exchange and analysis of data produced with non-proprietary chips:

- For example, the "Minimum Information About a Microarray Experiment" (MIAME) checklist helps define the level of detail that should exist and is being adopted by many journals as a requirement for the submission of papers incorporating microarray results. But MIAME does not describe the format for the information, so while many formats can support the MIAME requirements, as of 2007 no format permits verification of complete semantic compliance.
- The "MicroArray Quality Control (MAQC) Project" is being conducted by the US Food and Drug Administration (FDA) to develop standards and quality control metrics which will eventually allow the use of MicroArray data in drug discovery, clinical practice and regulatory decision-making.
- The MGED Society has developed standards for the representation of gene expression experiment results and relevant annotations.

Data Analysis

Microarray data sets are commonly very large, and analytical precision is influenced by a number of variables. Statistical challenges include taking into account effects of background noise and appropriate normalization of the data. Normalization methods may be suited to specific platforms and, in the case of commercial platforms, the analysis may be proprietary. Algorithms that affect statistical analysis include:

- Image analysis: gridding, spot recognition of the scanned image (segmentation algorithm), removal or marking of poor-quality and low-intensity features (called *flagging*).
- Data processing: background subtraction (based on global or local background), determination of spot intensities and intensity ratios, visualisation of data and log-transformation of ratios, global or local normalization of intensity ratios, and segmen-tation into different copy number regions using step detection algorithms.
- Class discovery analysis: This analytic approach, sometimes called unsupervised classification or knowledge discovery, tries to identify whether microarrays (objects, patients, mice, etc.) or genes cluster together in groups. Identifying naturally existing groups of objects (microarrays or genes) which cluster together can enable the discovery of new groups that otherwise were not previously known to exist. During knowledge discovery analysis, various unsupervised classification techniques can be employed with DNA microarray data to

identify novel clusters (classes) of arrays. This type of approach is not hypothesis-driven, but rather is based on iterative pattern recognition or statistical learning methods to find an "optimal" number of clusters in the data. Examples of unsupervised analyses methods include self-organizing maps, neural gas, k-means cluster analyses, hierarchical cluster analysis, Genomic Signal Processing based clustering and model-based cluster analysis. For some of these methods the user also has to define a distance measure between pairs of objects. Although the Pearson correlation coefficient is usually employed, several other measures have been proposed and evaluated in the literature. The input data used in class discovery analyses are commonly based on lists of genes having high informativeness (low noise) based on low values of the coefficient of variation or high values of Shannon entropy, etc. The determination of the most likely or optimal number of clusters obtained from an unsupervised analysis is called cluster validity. Some commonly used metrics for cluster validity are the silhouette index, Davies-Bouldin index, Dunn's index, or Hubert's statistic.

- Class prediction analysis: This approach, called supervised classification, establishes the basis for developing a predictive model into which future unknown test objects can be input in order to predict the most likely class membership of the test objects. Supervised analysis for class prediction involves use of techniques such as linear regression, k-nearest neighbor, learning vector quantization, decision tree analysis, random forests, naive Bayes, logistic regression, kernel regression, artificial neural networks, support vector machines, mixture of experts, and supervised neural gas. In addition, various metaheuristic methods are employed, such as genetic algorithms, covariance matrix self-adaptation, particle swarm optimization, and ant colony optimization. Input data for class prediction are usually based on filtered lists of genes which are predictive of class, determined using classical hypothesis tests (next section), Gini diversity index, or information gain (entropy).

- Hypothesis-driven statistical analysis: Identification of statistically significant changes in gene expression are commonly identified using the t-test, ANOVA, Bayesian method Mann–Whitney test methods tailored to microarray data sets, which take into account multiple comparisons or cluster analysis. These methods assess statistical power based on the variation present in the data and the number of experimental replicates, and can help minimize Type I and type II errors in the analyses.

- Dimensional reduction: Analysts often reduce the number of dimensions (genes) prior to data analysis. This may involve linear approaches such as principal components analysis (PCA), or non-linear manifold learning (distance metric learning) using kernel PCA, diffusion maps, Laplacian eigenmaps, local linear embedding, locally preserving projections, and Sammon's mapping.

- Network-based methods: Statistical methods that take the underlying structure of gene networks into account, representing either associative or causative interactions or dependencies among gene products. Weighted gene co-expression network analysis is widely used for identifying co-expression modules and intramodular hub genes. Modules may corresponds to cell types or pathways. Highly connected intramodular hubs best represent their respective modules.

Microarray data may require further processing aimed at reducing the dimensionality of the data to aid comprehension and more focused analysis. Other methods permit analysis of data consisting of a low number of biological or technical replicates; for example, the Local Pooled Error (LPE) test pools standard deviations of genes with similar expression levels in an effort to compensate for insufficient replication.

Annotation

The relation between a probe and the mRNA that it is expected to detect is not trivial. Some mRNAs may cross-hybridize probes in the array that are supposed to detect another mRNA. In addition, mRNAs may experience amplification bias that is sequence or molecule-specific. Thirdly, probes that are designed to detect the mRNA of a particular gene may be relying on genomic EST information that is incorrectly associated with that gene.

Data Warehousing

Microarray data was found to be more useful when compared to other similar datasets. The sheer volume of data, specialized formats (such as MIAME), and curation efforts associated with the datasets require specialized databases to store the data. A number of open-source data warehousing solutions, such as InterMine and BioMart, have been created for the specific purpose of integrating diverse biological datasets, and also support analysis.

Alternative Technologies

Advances in massively parallel sequencing has led to the development of RNA-Seq technology, that enables a whole transcriptome shotgun approach to characterize and quantify gene expression. Unlike microarrays, which need a reference genome and transcriptome to be available before the microarray itself can be designed, RNA-Seq can also be used for new model organisms whose genome has not been sequenced yet.

Multi-stranded DNA Microarray

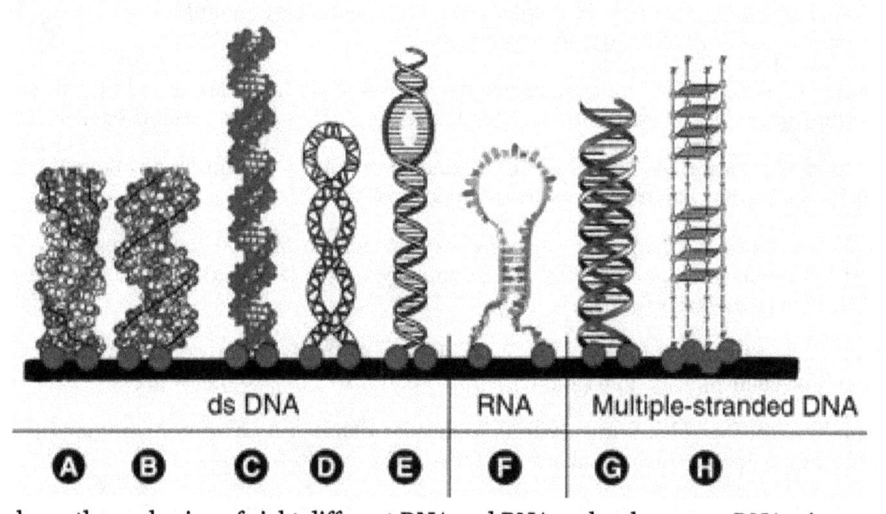

This figure shows the anchoring of eight different DNA and RNA molecules onto a DNA microarray surface.

Glossary

- An Array or slide is a collection of *features* spatially arranged in a two dimensional grid, arranged in columns and rows.
- Block or subarray: a group of spots, typically made in one print round; several subarrays/ blocks form an array.
- Case/control: an experimental design paradigm especially suited to the two-colour array system, in which a condition chosen as control (such as healthy tissue or state) is compared to an altered condition (such as a diseased tissue or state).
- Channel: the fluorescence output recorded in the scanner for an individual fluorophore and can even be ultraviolet.
- Dye flip or Dye swap or Fluor reversal: reciprocal labelling of DNA targets with the two dyes to account for dye bias in experiments.
- Scanner: an instrument used to detect and quantify the intensity of fluorescence of spots on a microarray slide, by selectively exciting fluorophores with a laser and measuring the fluorescence with a filter (optics) photomultiplier system.
- Spot or feature: a small area on an array slide that contains picomoles of specific DNA samples.
- For other relevant terms see:

  Glossary of gene expression terms

  Protocol (natural sciences)

References

- Patten CL, Glick BR, Pasternak J (2009). Molecular Biotechnology: Principles and Applications of Recombinant DNA. Washington, D.C: ASM Press. ISBN 978-1-55581-498-4.
- Russell DW, Sambrook J (2001). Molecular cloning: a laboratory manual. Cold Spring Harbor, N.Y: Cold Spring Harbor Laboratory. ISBN 978-0-87969-576-7.
- Peter Walter; Alberts, Bruce; Johnson, Alexander S.; Lewis, Julian; Raff, Martin C.; Roberts, Keith (2008). Molecular Biology of the Cell (5th edition, Extended version). New York: Garland Science. ISBN 0-8153-4111-3.
- Russell, David W.; Sambrook, Joseph (2001). Molecular cloning: a laboratory manual. Cold Spring Harbor, N.Y: Cold Spring Harbor Laboratory. ISBN 0-87969-576-5.
- Brondyk, W. H. (2009). "Chapter 11 Selecting an Appropriate Method for Expressing a Recombinant Protein". Methods in enzymology. Methods in Enzymology. 463: 131–147. doi:10.1016/S0076-6879(09)63011-1. ISBN 9780123745361. PMID 19892171.
- Bartlett, J. M. S.; Stirling, D. (2003). "A Short History of the Polymerase Chain Reaction". PCR Protocols. Methods in Molecular Biology. 226 (2nd ed.). pp. 3–6. doi:10.1385/1-59259-384-4:3. ISBN 1-59259-384-4.
- Joseph Sambrook & David W. Russel (2001). Molecular Cloning: A Laboratory Manual (3rd ed.). Cold Spring Harbor, N.Y.: Cold Spring Harbor Laboratory Press. ISBN 0-879-69576-5.
- Salis AD (2009). "Applications in Clinical Microbiology". Real-Time PCR: Current Technology and Applica-

tions. Caister Academic Press. ISBN 978-1-904455-39-4.

- Pierce KE & Wangh LJ (2007). "Linear-after-the-exponential polymerase chain reaction and allied technologies Real-time detection strategies for rapid, reliable diagnosis from single cells". Methods Mol Med. Methods in Molecular Medicine™. 132: 65–85. doi:10.1007/978-1-59745-298-4_7. ISBN 978-1-58829-578-1. PMID 17876077.
- Bartlett, J. M. S.; Stirling, D. (2003). "A Short History of the Polymerase Chain Reaction". PCR Protocols. Methods in Molecular Biology. 226 (2nd ed.). pp. 3–6. doi:10.1385/1-59259-384-4:3. ISBN 1-59259-384-4.
- Joseph Sambrook & David W. Russel (2001). Molecular Cloning: A Laboratory Manual (3rd ed.). Cold Spring Harbor, N.Y.: Cold Spring Harbor Laboratory Press. ISBN 0-879-69576-5.
- Pavlov AR, Pavlova NV, Kozyavkin SA, Slesarev AI (2006). "Thermostable DNA Polymerases for a Wide Spectrum of Applications: Comparison of a Robust Hybrid TopoTaq to other enzymes". In Kieleczawa J. DNA Sequencing II: Optimizing Preparation and Cleanup. Jones and Bartlett. pp. 241–257. ISBN 0-7637-3383-0.
- Salis AD (2009). "Applications in Clinical Microbiology". Real-Time PCR: Current Technology and Applications. Caister Academic Press. ISBN 978-1-904455-39-4.
- Pierce KE & Wangh LJ (2007). "Linear-after-the-exponential polymerase chain reaction and allied technologies Real-time detection strategies for rapid, reliable diagnosis from single cells". Methods Mol Med. Methods in Molecular Medicine™. 132: 65–85. doi:10.1007/978-1-59745-298-4_7. ISBN 978-1-58829-578-1. PMID 17876077.
- Pavlov AR, Pavlova NV, Kozyavkin SA, Slesarev AI (2006). "Thermostable DNA Polymerases for a Wide Spectrum of Applications: Comparison of a Robust Hybrid TopoTaq to other enzymes". In Kieleczawa J. DNA Sequencing II: Optimizing Preparation and Cleanup. Jones and Bartlett. pp. 241–257. ISBN 0-7637-3383-0.

Interdisciplinary Fields Related to Molecular Biology

The method used to imitate the behavior of molecules is molecular modelling whereas the testing of the properties of molecules in order to assemble better systems and processes is molecular engineering. This chapter is a compilation of fields like molecular modelling, molecular engineering, molecular genetics and molecular microbiology.

Molecular Modelling

Molecular modelling encompasses all theoretical methods and computational techniques used to model or mimic the behaviour of molecules. The techniques are used in the fields of computational chemistry, drug design, computational biology and materials science for studying molecular systems ranging from small chemical systems to large biological molecules and material assemblies. The simplest calculations can be performed by hand, but inevitably computers are required to perform molecular modelling of any reasonably sized system. The common feature of molecular modelling techniques is the atomistic level description of the molecular systems. This may include treating atoms as the smallest individual unit (the Molecular mechanics approach), or explicitly modeling electrons of each atom (the quantum chemistry approach).

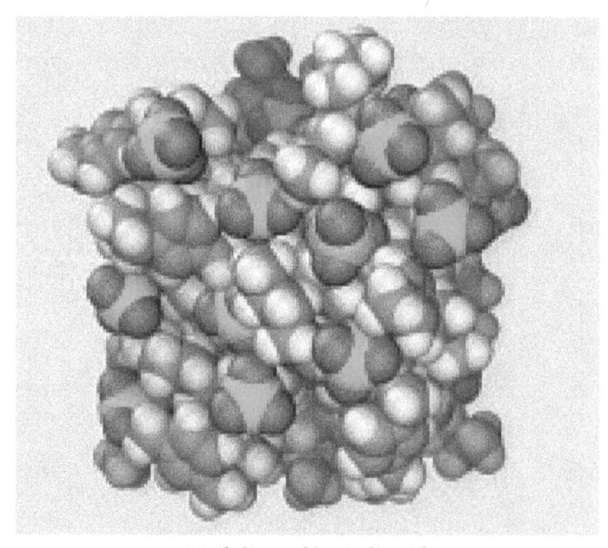

Modeling of ionic liquid

Molecular Mechanics

Molecular mechanics is one aspect of molecular modelling, as it refers to the use of classical mechanics/Newtonian mechanics to describe the physical basis behind the models. Molecular models typically describe atoms (nucleus and electrons collectively) as point charges with an associated mass. The interactions between neighbouring atoms are described by spring-like interactions (representing chemical bonds) and van der Waals forces. The Lennard-Jones potential is commonly used to describe van der Waals forces. The electrostatic interactions are computed based on Coulomb's law. Atoms are assigned coordinates in Cartesian space or in internal coordinates, and can also be assigned velocities in dynamical simulations. The atomic velocities are related to the temperature of the system, a macroscopic quantity. The collective mathematical expression is known as a potential function and is related to the system internal energy (U), a thermodynamic quantity equal to the sum of potential and kinetic energies. Methods which minimize the potential energy are known as energy minimization techniques (e.g., steepest descent and conjugate gradient), while methods that model the behaviour of the system with propagation of time are known as molecular dynamics.

$$E = E_{\text{bonds}} + E_{\text{angle}} + E_{\text{dihedral}} + E_{\text{non-bonded}}$$

$$E_{\text{non-bonded}} = E_{\text{electrostatic}} + E_{\text{van der Waals}}$$

This function, referred to as a potential function, computes the molecular potential energy as a sum of energy terms that describe the deviation of bond lengths, bond angles and torsion angles away from equilibrium values, plus terms for non-bonded pairs of atoms describing van der Waals and electrostatic interactions. The set of parameters consisting of equilibrium bond lengths, bond angles, partial charge values, force constants and van der Waals parameters are collectively known as a force field. Different implementations of molecular mechanics use different mathematical expressions and different parameters for the potential function. The common force fields in use today have been developed by using high level quantum calculations and/or fitting to experimental data. The technique known as energy minimization is used to find positions of zero gradient for all atoms, in other words, a local energy minimum. Lower energy states are more stable and are commonly investigated because of their role in chemical and biological processes. A molecular dynamics simulation, on the other hand, computes the behaviour of a system as a function of time. It involves solving Newton's laws of motion, principally the second law, $\mathbf{F} = m\mathbf{a}$. Integration of Newton's laws of motion, using different integration algorithms, leads to atomic trajectories in space and time. The force on an atom is defined as the negative gradient of the potential energy function. The energy minimization technique is useful for obtaining a static picture for comparing between states of similar systems, while molecular dynamics provides information about the dynamic processes with the intrinsic inclusion of temperature effects.

Variables

Molecules can be modeled either in vacuum or in the presence of a solvent such as water. Simulations of systems in vacuum are referred to as *gas-phase* simulations, while those that include the presence of solvent molecules are referred to as *explicit solvent* simulations. In another type of simulation, the effect of solvent is estimated using an empirical mathematical expression; these are known as *implicit solvation* simulations.

Coordinate Representations

Most force fields are distance-dependent, making the most convenient expression for these Cartesian coordinates. Yet the comparatively rigid nature of bonds which occur between specific atoms, and in essence, defines what we mean by the molecule itself, make an internal coordinate system the most logical representation. In some fields the IC representation (bond length, angle between bonds, and twist angle of the bond as shown in the figure) is known as the Z-matrix or torsion angle representation. Unfortunately, continuous motions in Cartesian space often require discontinuous angular branches in internal coordinates making it relatively hard to work with force fields in the internal coordinate representation and conversely a simple displacement of an atom in Cartesian space may not be a straight line trajectory due to the prohibitions of the interconnected bonds. Thus, it is very common for computational optimization programs to flip back and forth between representations during their iterations; this can dominate the calculation time of the potential itself and in long chain molecules introduce cumulative numerical inaccuracy. While all conversion algorithms produce mathematically identical results, they differ in speed and numerical accuracy. Currently, the fastest and most accurate torsion to Cartesian conversion is the Natural Extension Reference Frame (NERF) method.

Applications

Molecular modelling methods are now routinely used to investigate the structure, dynamics, surface properties and thermodynamics of inorganic, biological and polymeric systems. The types of biological activity that have been investigated using molecular modelling include protein folding, enzyme catalysis, protein stability, conformational changes associated with biomolecular function, and molecular recognition of proteins, DNA, and membrane complexes.

Molecular Engineering

Molecular engineering is an emerging field of study concerned with the design and testing of molecular properties, behavior and interactions in order to assemble better materials, systems, and processes for specific functions. This approach, in which observable properties of a macroscopic system are influenced by direct alteration of a molecular structure, falls into the broader category of "bottom-up" design.

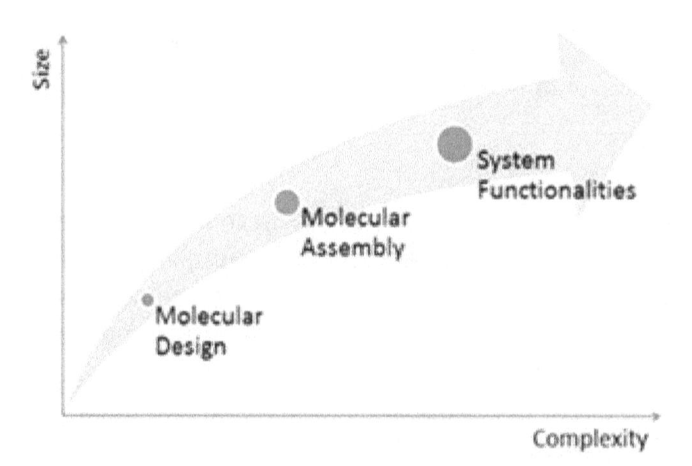

Molecular engineering deals with material development efforts in emerging technologies that require rigorous rational molecular design approaches towards systems of high complexity.

Molecular engineering is highly interdisciplinary by nature, encompassing aspects of chemical engineering, materials science, bioengineering, electrical engineering, physics, and chemistry. There is also considerable overlap with nanotechnology, in that both are concerned with the behavior of materials on the scale of nanometers or smaller. Given the highly fundamental nature of molecular interactions, there are a plethora of potential application areas, limited perhaps only by one's imagination and the laws of physics. However, some of the early successes of molecular engineering have come in the fields of immunotherapy, synthetic biology, and printable electronics (molecular engineering applications).

Molecular engineering is a dynamic and evolving field with complex target problems; breakthroughs require sophisticated and creative engineers who are conversant across disciplines. A rational engineering methodology that is based on molecular principles is in contrast to the widespread trial-and-error approaches common throughout engineering disciplines. Rather than relying on well-described but poorly-understood empirical correlations between the makeup of a system and its properties, a molecular design approach seeks to manipulate system properties directly using an understanding of their chemical and physical origins. This often gives rise to fundamentally new materials and systems, which are required to address outstanding needs in numerous fields, from energy to healthcare to electronics. Additionally, with the increased sophistication of technology, trial-and-error approaches are often costly and difficult, as it may be difficult to account for all relevant dependencies among variables in a complex system. Molecular engineering efforts may include computational tools, experimental methods, or a combination of both.

History

Molecular engineering was first mentioned in the research literature in 1956 by Arthur R. von Hippel, who defined it as "... a new mode of thinking about engineering problems. Instead of taking prefabricated materials and trying to devise engineering applications consistent with their macroscopic properties, one builds materials from their atoms and molecules for the purpose at hand." This concept was echoed in Richard Feynman's seminal 1959 lecture *There's Plenty of Room at the Bottom,* which is widely regarded as giving birth to some of the fundamental ideas of the field of nanotechnology. In spite of the early introduction of these concepts, it was not until the mid-1980s with the publication of *Engines of Creation: The Coming Era of Nanotechnology* by Drexler that the modern concepts of nano and molecular-scale science began to grow in the public consciousness.

The discovery of electrically-conductive properties in polyacetylene by Alan J. Heeger in 1977 effectively opened the field of organic electronics, which has proved foundational for many molecular engineering efforts. Design and optimization of these materials has led to a number of innovations including organic light-emitting diodes and flexible solar cells.

Applications

Molecular design has been an important element of many disciplines in academia, including bioengineering, chemical engineering, electrical engineering, materials science, mechanical engineering and chemistry. However, one of the on-going challenges is in bringing together the critical mass of manpower amongst disciplines to span the realm from design theory to materials production, and from device design to product development. Thus, while the concept of rational engineering of technology from the bottom-up is not new, it is still far from being widely translated into R&D efforts.

Molecular engineering is used in many industries. Some applications of technologies where molecular engineering plays a critical role:

Consumer Products

- Antibiotic surfaces
- Cosmetics (eg. rheological modification with small molecules in shampoo)
- Cleaning products (eg. nanosilver in laundry detergent)
- Consumer electronics (organic light-emitting diode displays (OLED)
- Electrochromic windows (eg. windows in Dreamliner 787)
- Zero emission vehicles (eg. advanced fuel cells/batteries)
- Self-cleaning surfaces (eg. super hydrophobic surface coatings)

Energy Harvesting and Storage

- Flow batteries - Synthesizing molecules for high-energy density electrolytes and highly-selective membranes in grid-scale energy storage systems.
- Lithium-ion batteries - Creating new molecules for use as electrode binders, electrolytes, electrolyte additives, or even for energy storage directly in order to improve energy density (using materials such as graphene, silicon nanorods, and lithium metal), power density, cycle life, and safety.
- Solar cells - Developing new materials for more efficient and cost-effective solar cells including organic, quantum dot or perovskite-based photovoltaics.
- Photocatalytic water splitting - Enhancing the production of hydrogen fuel using solar energy and advanced catalytic materials such as semiconductor nanoparticles

Environmental Engineering

- Water desalination (eg. new membranes for highly-efficient low-cost ion removal)
- Soil remediation (eg. catalytic nanoparticles that accelerate the degradation of long-lived soil contaminants such as chlorinated organic compounds)
- Carbon sequestration (eg. new materials for CO_2 adsorption)
- Wastewater treatment

Immunotherapy

Synthetic Biology

- CRISPR - Faster and more efficient gene editing technique

- Gene delivery/gene therapy - Designing molecules to deliver modified or new genes into cells of live organisms to cure genetic disorders
- Metabolic engineering - Modifying metabolism of organisms to optimize production of chemicals (eg. synthetic genomics)
- Protein engineering - Altering structure of existing proteins to enable specific new functions, or the creation of fully artificial proteins

Techniques and Instruments Used

Molecular engineers utilize sophisticated tools and instruments to make and analyze the interactions of molecules and the surfaces of materials at the molecular and nano-scale. The complexity of molecules being introduced at the surface is increasing, and the techniques used to analyze surface characteristics at the molecular level are ever-changing and improving. Meantime, advancements in high performance computing have greatly expanded the use of computer simulation in the study of molecular scale systems.

Computational and Theoretical Approaches

- Computational chemistry
- High performance computing
- Molecular dynamics
- Molecular modeling
- Statistical mechanics
- Theoretical chemistry

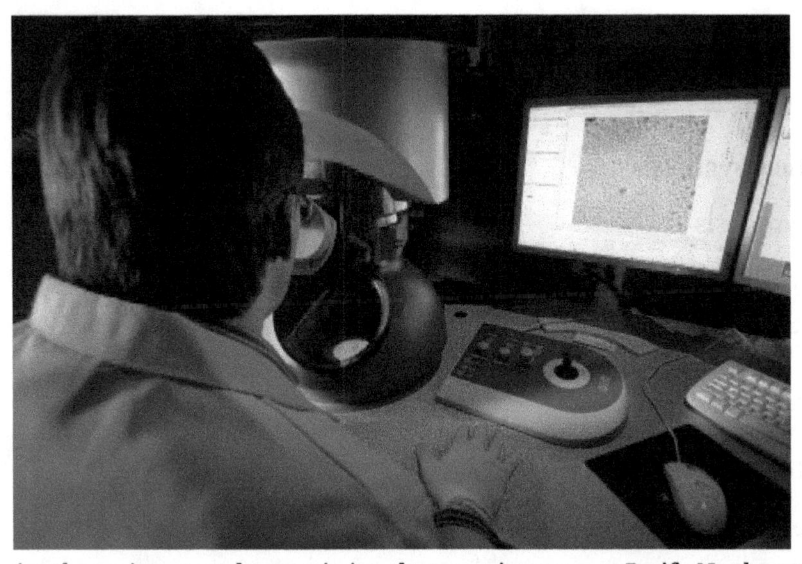

An EMSL scientist using the environmental transmission electron microscope at Pacific Northwest National Laboratory. The ETEM provides in situ capabilities that enable atomic-resolution imaging and spectroscopic studies of materials under dynamic operating conditions. In contrast to traditional operation of TEM under high vacuum, EMSL's ETEM uniquely allows imaging within high-temperature and gas environments.

Microscopy

- Atomic Force Microscopy (AFM)
- Scanning Electron Microscopy (SEM)
- Transmission Electron Microscopy (TEM)

Molecular Characterization

- Dynamic light scattering (DLS)
- Matrix-assisted laser desorption/ionization (MALDI) spectrocosopy
- Nuclear magnetic resonance (NMR) spectroscopy
- Size exclusion chromatography (SEC)

Spectroscopy

- Ellipsometry
- 2D X-Ray Diffraction (XRD)
- Raman Spectroscopy/Microscopy

Surface Science

- Glow Discharge Optical Emission Spectrometry
- Time of Flight-Secondary Ion Mass Spectrometry (ToF-SIMS)
- X-Ray Photoelectron Spectroscopy (XPS)

Synthetic Methods

- DNA synthesis
- Nanoparticle synthesis
- Organic synthesis
- Peptide synthesis
- Polymer synthesis

Other Tools

- Focused Ion Beam (FIB)
- Profilometer
- UV Photoelectron Spectroscopy (UPS)
- Vibrational Sum Frequency Generation

Research / Education

At least three universities offer graduate degrees dedicated to molecular engineering: the University of Chicago, the University of Washington, and Kyoto University. These programs are interdisciplinary institutes with faculty from several research areas.

The academic journal Molecular Systems Design & Engineering publishes research from a wide variety of subject areas that demonstrates "a molecular design or optimisation strategy targeting specific systems functionality and performance."

Molecular Genetics

Molecular genetics is the field of biology and genetics that studies the structure and function of genes at a molecular level. The study of chromosomes and gene expression of an organism can give insight into heredity, genetic variation, and mutations. This is useful in the study of developmental biology and in understanding and treating genetic diseases.

Techniques in Molecular Genetics

Amplification

Gene amplification is a procedure in which a certain gene or DNA sequence is replicated many times in a process called DNA replication.

Polymerase Chain Reaction

> The main genetic components of the polymerase chain reaction (PCR) are DNA nucleotides, template DNA, primers and Taq polymerase. DNA nucleotides make up the DNA template strand for the specific sequence being amplified, primers are short strands of complementary nucleotides where DNA replication starts, and Taq polymerase is a heat stable enzyme that jump-starts the production of new DNA at the high temperatures needed for reaction.

Cloning DNA in Bacteria

> Cloning is the process of creating many identical copies of a sequence of DNA. The target DNA sequence is inserted into a cloning vector. Because this vector originates from a self-replicating virus, plasmid, or higher organism cell when the appropriate size DNA is inserted the "target and vector DNA fragments are then ligated" and create a recombinant DNA molecule.

The recombinant DNA molecules are then put into a bacterial strain (usually *E. coli*) which produces several identical copies by transformation. Transformation is the DNA uptake mechanism possessed by bacteria. However, only one recombinant DNA molecule can be cloned within a single bacteria cell, so each clone is of just one DNA insert.

Separation and Detection

In separation and detection DNA and mRNA are isolated from cells and then detected simply by the isolation. Cell cultures are also grown to provide a constant supply of cells ready for isolation.

Cell cultures

A cell culture for molecular genetics is a culture that is grown in artificial conditions. Some cell types grow well in cultures such as skin cells, but other cells are not as productive in cultures. There are different techniques for each type of cell, some only recently being found to foster growth in stem and nerve cells. Cultures for molecular genetics are frozen in order to preserve all copies of the gene specimen and thawed only when needed. This allows for a steady supply of cells.

DNA isolation

DNA isolation extracts DNA from a cell in a pure form. First, the DNA is separated from cellular components such as proteins, RNA, and lipids. This is done by placing the chosen cells in a tube with a solution that mechanically, chemically, breaks the cells open. This solution contains enzymes, chemicals, and salts that breaks down the cells except for the DNA. It contains enzymes to dissolve proteins, chemicals to destroy all RNA present, and salts to help pull DNA out of the solution. Next, the DNA is separated from the solution by being spun in a centrifuge, which allows the DNA to collect in the bottom of the tube. After this cycle in the centrifuge the solution is poured off and the DNA is resuspended in a second solution that makes the DNA easy to work with in the future. This results in a concentrated DNA sample that contains thousands of copies of each gene. For large scale projects such as sequencing the human genome, all this work is done by robots.

mRNA isolation

Expressed DNA that codes for the synthesis of a protein is the final goal for scientists and this expressed DNA is obtained by isolating mRNA (Messenger RNA).

First, laboratories use a normal cellular modification of mRNA that adds up to 200 adenine nucleotides to the end of the molecule (poly(A) tail). Once this has been added, the cell is ruptured and its cell contents are exposed to synthetic beads that are coated with thymine string nucleotides. Because Adenine and Thymine pair together in DNA, the poly(A) tail and synthetic beads are attracted to one another, and once they bind in this process the cell components can be washed away without removing the mRNA. Once the mRNA has been isolated, reverse transcriptase is employed to convert it to single-stranded DNA, from which a stable double-stranded DNA is produced using DNA polymerase. Complementary DNA (cDNA) is much more stable than mRNA and so, once the double-stranded DNA has been produced it represents the expressed DNA sequence scientists look for.

Genetic Screens

This technique is used to identify which genes or genetic mutations produce a certain phenotype. A mutagen is very often used to accelerate this process. Once mutants have been isolated, the mutated genes can be molecularly identified.

Forward saturation genetics is a method for treating organisms with a mutagen, then screens the organism's offspring for particular phenotypes. This type of genetic screening is used to find and identify all the genes involved in a trait.

Reverse genetics

> Reverse genetics determines the phenotype that results from a specifically engineered gene. In some organisms, such as yeast and mice, it is possible to induce the deletion of a particular gene, creating what's known as a gene "knockout" - the laboratory origin of so-called "knockout mice" for further study. In other words this process involves the creation of transgenic organisms that do not express a gene of interest. Alternative methods of reverse genetic research include the random induction of DNA deletions and subsequent selection for deletions in a gene of interest, as well as the application of RNA interference.

Gene Therapy

A mutation in a gene can cause encoded proteins and the cells that rely on those proteins to malfunction. Conditions related to gene mutations are called genetic disorders. However, altering a patient's genes can sometimes be used to treat or cure a disease as well. Gene therapy can be used to replace a mutated gene with the correct copy of the gene, to inactivate or knockout the expression of a malfunctioning gene, or to introduce a foreign gene to the body to help fight disease. Major diseases that can be treated with gene therapy include viral infections, cancers, and inherited disorders, including immune system disorders.

Gene therapy delivers a copy of the missing, mutated, or desired gene via a modified virus or vector to the patient's target cells so that a functional form of the protein can then be produced and incorporated into the body. These vectors are often siRNA. Treatment can be either in vivo or ex vivo. The therapy has to be repeated several times for the infected patient to continually be relieved, as repeated cell division and cell death slowly randomizes the body's ratio of functional-to-mutant genes. Gene therapy is an appealing alternative to some drug-based approaches, because gene therapy repairs the underlying genetic defect using the patients own cells with minimal side effects. Gene therapies are still in development and mostly used in research settings. All experiments and products are controlled by the U.S. FDA and the NIH.

Classical gene therapies usually require efficient transfer of cloned genes into the disease cells so that the introduced genes are expressed at sufficiently high levels to change the patient's physiology. There are several different physicochemical and biological methods that can be used to transfer genes into human cells. The size of the DNA fragments that can be transferred is very limited, and often the transferred gene is not a conventional gene. Horizontal gene transfer is the transfer of genetic material from one cell to another that is not its offspring. Artificial horizontal gene transfer is a form of genetic engineering.

The Human Genome Project

The Human Genome Project is a molecular genetics project that began in the 1990s and was projected to take fifteen years to complete. However, because of technological advances the progress of the project was advanced and the project finished in 2003, taking only thirteen years. The project was started by the U.S. Department of Energy and the National Institutes of Health in an effort to reach six set goals. These goals included:

1. identifying 20,000 to 25,000 genes in human DNA (although initial estimates were approximately 100,000 genes),
2. determining sequences of chemical base pairs in human DNA,
3. storing all found information into databases,
4. improving the tools used for data analysis,
5. transferring technologies to private sectors, and
6. addressing the ethical, legal, and social issues (ELSI) that may arise from the projects.

The project was worked on by eighteen different countries including the United States, Japan, France, Germany, and the United Kingdom. The collaborative effort resulted in the discovery of the many benefits of molecular genetics. Discoveries such as molecular medicine, new energy sources and environmental applications, DNA forensics, and livestock breeding, are only a few of the benefits that molecular genetics can provide.

Molecular Microbiology

Molecular microbiology is the branch of microbiology devoted to the study of the molecular basis of the physiological processes that occur in microorganisms.

Bacteria

Mainly because of their relative simplicity, ease of manipulation and growth in vitro, and importance in medicine, bacteria were instrumental in the development of molecular biology. The complete genome sequence for a large number of bacterial species is now available. A list of sequenced prokaryotic genomes is available. Molecular microbiology techniques are currently being used in the development of new genetically engineered vaccines, in bioremediation, biotechnology, food microbiology, probiotic research, antibacterial development and environmental microbiology.

Many bacteria have become model organisms for molecular studies.

Molecular techniques have had a direct influence on the clinical practice of medical microbiology. In many cases where traditional phenotypic methods of microbial identification and typing are insufficient or time-consuming, molecular techniques can provide rapid and accurate data, potentially improving clinical outcomes. Specific examples include:

- 16s rRNA sequencing to provide bacterial identifications
- Pulsed Field Gel Electrophoresis for strain typing of epidemiologically related organisms.
- Direct detection of genes related to resistance mechanisms, such as mecA gene in Staphylococcus aureus

Signaling

Bacteria possess diverse proteins and RNA that can sense changes to their intracellular and extra-cellular environment. The signals received by these macromolecules are transmitted to key genes or proteins, which alter their activities to suit the new conditions.

Viruses

Viruses are important pathogens of humans and animals. Their genomes are relatively small. For these reasons they were among the first organisms to be fully sequenced. The complete DNA sequence of the Epstein-Barr virus was completed in 1984. Bluetongue virus (BTV) has been in the forefront of molecular studies for last three decades and now represents one of the best understood viruses at the molecular and structural levels. Other viruses such as Papillomavirus, Coronavirus, Caliciviruses, Paramyxoviruses and Influenza virus have also been extensively studied at the molecular level.

Bacterial viruses, or bacteriophages, are estimated to be the most widely distributed and diverse entities in the biosphere. Bacteriophages, or "phage", have been fundamental in the development of the science of molecular biology and became "model organisms" for probing the basic chemistry of life. The first DNA-genome project to be completed was the phage Φ-X174 in 1977. Φ29 phage, a phage of Bacillus, is a paradigm for the study of several molecular mechanisms of general biological processes, including DNA replication and regulation of transcription.

Gene Therapy

Some viruses are used as vectors for gene therapy. Virus vectors have been developed that mediate stable genetic modification of treated cells by chromosomal integration of the transferred vector genomes. Gammaretroviral and lentiviral vectors, for example, can be utilized in clinical gene therapy aimed at the long-term correction of genetic defects, e.g., in stem and progenitor cells. Gammaretroviral and lentiviral vectors have so far been used in more than 300 clinical trials, addressing treatment options for various diseases.

Fungi

Yeasts and molds are eukaryotic microorganisms classified in the kingdom Fungi.

Technology

Polymerase chain reaction (PCR) is used in microbiology to amplify (replicate many times) a single DNA sequence. If required, the sequence can also be altered in predetermined ways. Quantitative PCR is used for the rapid detection of microorganisms and is currently employed in diagnostic clinical microbiology laboratories, environmental analysis, food microbiology, and many other fields. The closely related technique of quantitative PCR permits the quantitative measurement of DNA or RNA molecules and is used to estimate the densities of the reference pathogens in food, water and environmental samples. Quantitative PCR provides both specificity and quantification of target microorganisms.

Gel electrophoresis is used routinely in microbiology to separate DNA, RNA, or protein molecules using an electric field by virtue of their size, shape or electric charge.

Southern blotting, northern blotting, western blotting and Eastern blotting are molecular techniques for detecting the presence of microbial DNA sequences (Southern), RNA sequences (northern), protein molecules (western) or protein modifications (Eastern).

DNA microarrays are used in microbiology as the modern alternative to the "blotting" techniques. Microarrays permit the exploration of thousands of sequences at one time. This technique is used in molecular microbiology to detect the presence of pathogens in a sample (air, water, organ tissue, etc.). It is also used to determine the genetic differences between two microbial strains.

DNA sequencing and genomics have been used for many decades in molecular microbiology studies. Due to their relatively small size, viral genomes were the first to be completely analysed by DNA sequencing. A huge range of sequence and genomic data is now available for a number of species and strains of microorganisms.

RNA interference (RNAi) was discovered as a cellular gene regulation mechanism in 1998, but several RNAi-based applications for gene silencing have already made it into clinical trials. RNA interference (RNAi) technology has formed the basis of novel tools for biological research and drug discovery.

References

- Diaz E (editor). (2008). Microbial Biodegradation: Genomics and Molecular Biology. Caister Academic Press. ISBN 978-1-904455-17-2.
- Fratamico PM and Bayles DO (editor). (2005). Foodborne Pathogens: Microbiology and Molecular Biology. Caister Academic Press. ISBN 978-1-904455-00-4.
- Mayo, B; van Sinderen, D (editor) (2010). Bifidobacteria: Genomics and Molecular Aspects. Caister Academic Press. ISBN 978-1-904455-68-4.
- Miller, AA; Miller, PF (editor) (2011). Emerging Trends in Antibacterial Discovery: Answering the Call to Arms. Caister Academic Press. ISBN 978-1-904455-89-9.
- Sen, K; Ashbolt, NK (editor) (2010). Environmental Microbiology: Current Technology and Water Applications. Caister Academic Press. ISBN 978-1-904455-70-7.
- Roy P (2008). "Molecular Dissection of Bluetongue Virus". Animal Viruses: Molecular Biology. Caister Academic Press. ISBN 978-1-904455-22-6.
- Campo MS (editor). (2006). Papillomavirus Research: From Natural History To Vaccines and Beyond. Caister Academic Press. ISBN 978-1-904455-04-2.
- Mc Grath S and van Sinderen D (editors). (2007). Bacteriophage: Genetics and Molecular Biology. Caister Academic Press. ISBN 978-1-904455-14-1.
- Kurth, R; Bannert, N (editors) (2010). Retroviruses: Molecular Biology, Genomics and Pathogenesis. Caister Academic Press. ISBN 978-1-904455-55-4.
- Herold, KE; Rasooly, A (editor) (2009). Lab-on-a-Chip Technology: Biomolecular Separation and Analysis. Caister Academic Press. ISBN 978-1-904455-47-9.
- Morris, KV (editor) (2008). RNA and the Regulation of Gene Expression: A Hidden Layer of Complexity. Caister Academic Press. ISBN 978-1-904455-25-7.

5

Applications of Molecular Biology

This chapter will provide an integrated understanding of the applications of molecular biology, with a brief explanation on molecular medicine, bioinformatics and central dogma of molecular biology. This chapter helps the reader to develop a better understanding on the applications of molecular biology.

Molecular Medicine

Molecular medicine is a broad field, where physical, chemical, biological and medical techniques are used to describe molecular structures and mechanisms, identify fundamental molecular and genetic errors of disease, and to develop molecular interventions to correct them. The molecular medicine perspective emphasizes cellular and molecular phenomena and interventions rather than the previous conceptual and observational focus on patients and their organs.

In November 1949, with the seminal paper, "Sickle Cell Anemia, a Molecular Disease", in *Science* magazine, Linus Pauling, Harvey Itano and their collaborators laid the groundwork for establishing the field of molecular medicine. In 1956, Roger J. Williams wrote *Biochemical Individuality*, a prescient book about genetics, prevention and treatment of disease on a molecular basis, and nutrition which is now variously referred to as individualized medicine and orthomolecular medicine. Another paper in *Science* by Pauling in 1968, introduced and defined this view of molecular medicine that focuses on natural and nutritional substances used for treatment and prevention.

Published research and progress was slow until the 1970s' "biological revolution" that introduced many new techniques and commercial applications.

Molecular medicine is a new scientific discipline in European universities. Combining contemporary medical studies with the field of biochemistry, it offers a bridge between the two subjects. At present only a handful of universities offer the course to undergraduates. With a degree in this discipline the graduate is able to pursue a career in medical sciences, scientific research, laboratory work and postgraduate medical degrees.

Subjects

Core subjects are similar to biochemistry courses and typically include gene expression, research methods, proteins, cancer research, immunology, biotechnology and many more besides. In some universities molecular medicine is combined with another discipline such as chemistry, functioning as an additional study to enrich the undergraduate program.

Bioinformatics

Bioinformatics is an interdisciplinary field that develops methods and software tools for understanding biological data. As an interdisciplinary field of science, bioinformatics combines computer science, statistics, mathematics, and engineering to analyze and interpret biological data. Bioinformatics has been used for *in silico* analyses of biological queries using mathematical and statistical techniques.

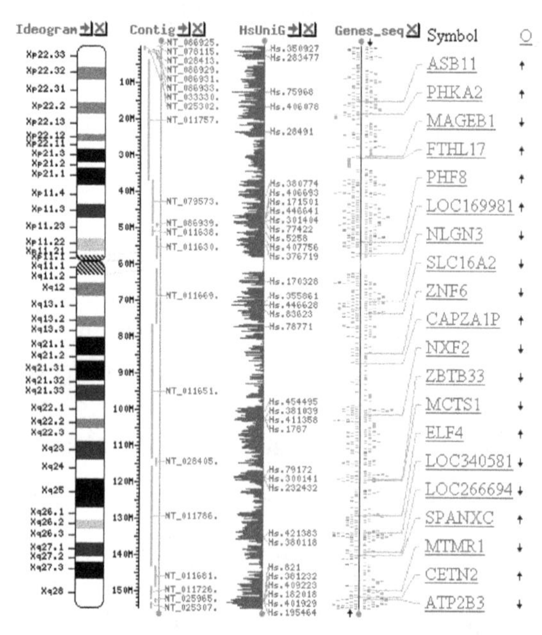

Map of the human X chromosome (from the National Center for Biotechnology Information website).

Bioinformatics is both an umbrella term for the body of biological studies that use computer programming as part of their methodology, as well as a reference to specific analysis "pipelines" that are repeatedly used, particularly in the field of genomics. Common uses of bioinformatics include the identification of candidate genes and nucleotides (SNPs). Often, such identification is made with the aim of better understanding the genetic basis of disease, unique adaptations, desirable properties (esp. in agricultural species), or differences between populations. In a less formal way, bioinformatics also tries to understand the organisational principles within nucleic acid and protein sequences.

Introduction

Bioinformatics has become an important part of many areas of biology. In experimental molecular biology, bioinformatics techniques such as image and signal processing allow extraction of useful results from large amounts of raw data. In the field of genetics and genomics, it aids in sequencing and annotating genomes and their observed mutations. It plays a role in the text mining of biological literature and the development of biological and gene ontologies to organize and query biological data. It also plays a role in the analysis of gene and protein expression and regulation. Bioinformatics tools aid in the comparison of genetic and genomic data and more generally in the understanding of evolutionary aspects of molecular biology. At a more integrative level, it helps analyze and catalogue the biological pathways and networks that are an important part of systems biology. In structural biology, it aids in the simulation and modeling of DNA, RNA, proteins as well as biomolecular interactions.

History

Historically, the term *bioinformatics* did not mean what it means today. Paulien Hogeweg and Ben Hesper coined it in 1970 to refer to the study of information processes in biotic systems. This definition placed bioinformatics as a field parallel to biophysics (the study of physical processes in biological systems) or biochemistry (the study of chemical processes in biological systems).

Sequences

5'ATGACGTGGGGA3'
3'TACTGCACCCCT5'

Sequences of genetic material are frequently used in bioinformatics and are easier to manage using computers than manually.

Computers became essential in molecular biology when protein sequences became available after Frederick Sanger determined the sequence of insulin in the early 1950s. Comparing multiple sequences manually turned out to be impractical. A pioneer in the field was Margaret Oakley Dayhoff, who has been hailed by David Lipman, director of the National Center for Biotechnology Information, as the "mother and father of bioinformatics." Dayhoff compiled one of the first protein sequence databases, initially published as books and pioneered methods of sequence alignment and molecular evolution. Another early contributor to bioinformatics was Elvin A. Kabat, who pioneered biological sequence analysis in 1970 with his comprehensive volumes of antibody sequences released with Tai Te Wu between 1980 and 1991.

Goals

To study how normal cellular activities are altered in different disease states, the biological data must be combined to form a comprehensive picture of these activities. Therefore, the field of bioinformatics has evolved such that the most pressing task now involves the analysis and interpretation of various types of data. This includes nucleotide and amino acid sequences, protein domains, and protein structures. The actual process of analyzing and interpreting data is referred to as computational biology. Important sub-disciplines within bioinformatics and computational biology include:

- Development and implementation of computer programs that enable efficient access to, use and management of, various types of information
- Development of new algorithms (mathematical formulas) and statistical measures that assess relationships among members of large data sets. For example, there are methods to locate a gene within a sequence, to predict protein structure and/or function, and to cluster protein sequences into families of related sequences.

The primary goal of bioinformatics is to increase the understanding of biological processes. What sets it apart from other approaches, however, is its focus on developing and applying computationally intensive techniques to achieve this goal. Examples include: pattern recognition, data mining, machine learning algorithms, and visualization. Major research efforts in the field include sequence alignment, gene finding, genome assembly, drug design, drug discovery, protein structure alignment, protein structure prediction, prediction of gene expression and protein–protein interactions, genome-wide association studies, the modeling of evolution and cell division/mitosis.

Bioinformatics now entails the creation and advancement of databases, algorithms, computational and statistical techniques, and theory to solve formal and practical problems arising from the management and analysis of biological data.

Over the past few decades, rapid developments in genomic and other molecular research technologies and developments in information technologies have combined to produce a tremendous amount of information related to molecular biology. Bioinformatics is the name given to these mathematical and computing approaches used to glean understanding of biological processes.

Common activities in bioinformatics include mapping and analyzing DNA and protein sequences, aligning DNA and protein sequences to compare them, and creating and viewing 3-D models of protein structures.

Relation to Other Fields

Bioinformatics is a science field that is similar to but distinct from biological computation and computational biology. Biological computation uses bioengineering and biology to build biological computers, whereas bioinformatics uses computation to better understand biology. Bioinformatics and computational biology have similar aims and approaches, but they differ in scale: bioinformatics organizes and analyzes basic biological data, whereas computational biology builds theoretical models of biological systems, just as mathematical biology does with mathematical models.

Analyzing biological data to produce meaningful information involves writing and running software programs that use algorithms from graph theory, artificial intelligence, soft computing, data mining, image processing, and computer simulation. The algorithms in turn depend on theoretical foundations such as discrete mathematics, control theory, system theory, information theory, and statistics.

Sequence Analysis

```
A5ASC3.1   14 SIKLWPPSQTTRLLLVERMANNLST..PSIFTRK..YGSLSKEEARENAKQIEEVACSTANQ.....HYEKEPDGDGGSAVQLYAKECSKLILEVLK 101
B4F917.1   13 SIKLWPPSESTRIMLVDRMTNNLST..ESIFSRK..YRLLGKQEAHENAKTIEELCFALADE.....HFREEPDGDGSSAVQLYAKETSKMMLEVLK 100
A9S1V2.1   23 VFKLWPPSQGTREAVRQKMALKLSS..ACFESQS..FARIELADAQEHARAIEEVAFGAAQE......ADSGGDKTGSAVVMVYAKHASKLMLETLR 109
B9GSN7.1   13 SVKLWPPGQSTRLMLVERMTKNFIT..PSFISRK..YGLLSKEEAEEDAKKIEEVAFAAANQ.....HYEKQPDGDGSSAVQIYAKESSRLMLEVLK 100
Q8H056.1   30 SFSIWPPTQRTRDAVVRRLVDTLGG..DTILCKR..YGAVPAADAEPAARGIEAEAFDAAAA..SGEAAATASVEEGIKALQLYSKEVSRRLLDFVK 120
Q0D4Z3.2   44 SLSIWPPSQRTRDAVVRRLVQTLVA..PSILSQR..YGAVPEAEAGRAAAAVEAEAYAAVTES.SSAAAAPASVEDGIEVLQAYSKEVSRRLLELAK 135
B9MVW8.1   56 SFSIWPPTQRTRDAIISRLIETLST..TSVLSKR..YGTIPKEEASEASRRIEEEAFSGAST.......VASSEKDGLEVLQLYSKEISKRMLETVK 141
Q0IYC5.1   29 SFAVWPPTRRTRDAVVRRLVAVLSGDTTTALRKRYRYGAVPAADAERAARAVEAQAFDAASA....SSSSSSSVEDGIETLQLYSREVSNRLLAFVR 121
A9NW46.1   13 SIKLWPPSESTRLMLVERMTDNLSS..VSFFSRK..YGLLSKEEAAENAKRIEETAFLAAND.....HEAKEPNLDDSSVVQFYAREASKLMLEALK 100
Q9C500.1   57 SLRIWPPTQKTRDAVLNRLIETLST..ESILSKR..YGTLKSDDATTVAKLIEEEAYGVASN.......AVSSDDDGIKILELYSKEISKRMLESVK 142
Q2HRI7.1   25 NYSIWPPKQRTRDAVKNRLIETLST..PSVLTKR..YGTMSADEASAAAIQIEDEAFSVANA.......SSSTSNDNVTILEVYSKEISKRMIETVK 110
Q9M7N3.1   28 SFKIWPPTQRTREAVVRRLVETLTS..QSVLSKR..YGVIPEEDATSAARIIEEEAFSVASV.ASAASTGGRPEDEWIEVLHIYSQEIXQRVVESAK 119
Q9M7N6.1   25 SFSIWPPTQRTRDAVINRLIESLST..PSILSKR..YGTLPQDEASETARLIEEEAFAAAGS.......TASDADDGIEILQVYSKEISKRMIDTVK 110
Q9LE82.1   14 SVKMWPPSKSTRLMLVERMTKNITT..PSIFSRK..YGLLSVEEAEQDAKRIEDLAFATANK.....HFQNEPDGDGTSAVHVYAKESSKLMLDVIK 101
Q9M651.2   13 SIKLWPPSLPTRKALIERITNNFSS..KTIFTEK..YGSLTKDQATENAKRIEDIAFSTANQ.....QFEREPDGDGGSAVQLYAKECSKLILEVLK 100
B9R748.1   48 SLSIWPPTQRTRDAVITRLIETLSS..PSVLSKR..YGTISHDEAESAARRIEDEAFGVANT.......ATSAEDDGLEILQLYSKEISRRMLDTVK 133
```

The sequences of different genes or proteins may be aligned side-by-side to measure their similarity. This alignment compares protein sequences containing WPP domains.

Since the Phage Φ-X174 was sequenced in 1977, the DNA sequences of thousands of organisms have been decoded and stored in databases. This sequence information is analyzed to determine genes that encode proteins, RNA genes, regulatory sequences, structural motifs, and repetitive sequences. A comparison of genes within a species or between different species can show similarities between protein functions, or relations between species (the use of molecular systematics to construct phylogenetic trees). With the growing amount of data, it long ago became impractical to analyze DNA sequences manually. Today, computer programs such as BLAST are used daily to search sequences from more than 260 000 organisms, containing over 190 billion nucleotides. These programs can compensate for mutations (exchanged, deleted or inserted bases) in the DNA sequence, to identify sequences that are related, but not identical. A variant of this sequence alignment is used in the sequencing process itself. The so-called shotgun sequencing technique (which was used, for example, by The Institute for Genomic Research (TIGR) to sequence the first bacterial genome, *Haemophilus influenzae*) does not produce entire chromosomes. Instead it generates the sequences of many thousands of small DNA fragments (ranging from 35 to 900 nucleotides long, depending on the sequencing technology). The ends of these fragments overlap and, when aligned properly by a genome assembly program, can be used to reconstruct the complete genome. Shotgun sequencing yields sequence data quickly, but the task of assembling the fragments can be quite complicated for larger genomes. For a genome as large as the human genome, it may take many days of CPU time on large-memory, multiprocessor computers to assemble the fragments, and the resulting assembly usually contains numerous gaps that must be filled in later. Shotgun sequencing is the method of choice for virtually all genomes sequenced today, and genome assembly algorithms are a critical area of bioinformatics research.

Following the goals that the Human Genome Project left to achieve after its closure in 2003, a new project developed by the National Human Genome Research Institute in the U.S appeared. The so-called ENCODE project is a collaborative data collection of the functional elements of the human genome that uses next-generation DNA-sequencing technologies and genomic tiling arrays, technologies able to generate automatically large amounts of data with lower research costs but with the same quality and viability.

Another aspect of bioinformatics in sequence analysis is annotation. This involves computational gene finding to search for protein-coding genes, RNA genes, and other functional sequences within a genome. Not all of the nucleotides within a genome are part of genes. Within the genomes of higher organisms, large parts of the DNA do not serve any obvious purpose.

Genome Annotation

In the context of genomics, annotation is the process of marking the genes and other biological features in a DNA sequence. This process needs to be automated because most genomes are too large to annotate by hand, not to mention the desire to annotate as many genomes as possible, as the rate of sequencing has ceased to pose a bottleneck. Annotation is made possible by the fact that genes have recognisable start and stop regions, although the exact sequence found in these regions can vary between genes.

The first genome annotation software system was designed in 1995 by Owen White, who was part of the team at The Institute for Genomic Research that sequenced and analyzed the first genome of a free-living organism to be decoded, the bacterium *Haemophilus influenzae*. White built a software system to find the genes (fragments of genomic sequence that encode proteins), the transfer RNAs, and to make initial assignments of function to those genes. Most current genome annotation systems work similarly, but the programs available for analysis of genomic DNA, such as the GeneMark program trained and used to find protein-coding genes in *Haemophilus influenzae*, are constantly changing and improving.

Computational Evolutionary Biology

Evolutionary biology is the study of the origin and descent of species, as well as their change over time. Informatics has assisted evolutionary biologists by enabling researchers to:

- trace the evolution of a large number of organisms by measuring changes in their DNA, rather than through physical taxonomy or physiological observations alone,
- more recently, compare entire genomes, which permits the study of more complex evolutionary events, such as gene duplication, horizontal gene transfer, and the prediction of factors important in bacterial speciation,
- build complex computational models of populations to predict the outcome of the system over time
- track and share information on an increasingly large number of species and organisms

Future work endeavours to reconstruct the now more complex tree of life.

The area of research within computer science that uses genetic algorithms is sometimes confused with computational evolutionary biology, but the two areas are not necessarily related.

Comparative Genomics

The core of comparative genome analysis is the establishment of the correspondence between genes (orthology analysis) or other genomic features in different organisms. It is these intergenomic maps that make it possible to trace the evolutionary processes responsible for the divergence of two genomes. A multitude of evolutionary events acting at various organizational levels shape genome evolution. At the lowest level, point mutations affect individual nucleotides. At a higher level, large chromosomal segments undergo duplication, lateral transfer, inversion, transposition, deletion and insertion. Ultimately, whole genomes are involved in processes of hybridization, polyploidization and endosymbiosis, often leading to rapid speciation. The complexity of genome evolution poses many exciting challenges to developers of mathematical models and algorithms, who have recourse to a spectra of algorithmic, statistical and mathematical techniques, ranging from exact, heuristics, fixed parameter and approximation algorithms for problems based on parsimony models to Markov Chain Monte Carlo algorithms for Bayesian analysis of problems based on probabilistic models.

Many of these studies are based on the homology detection and protein families computation.

Pan Genomics

Pan genomics is a concept introduced in 2005 by Tettelin and Medini which eventually took root in bioinformatics. Pan genome is the complete gene repertoire of a particular taxonomic group: although initially applied to closely related strains of a species, it can be applied to a larger context like genus, phylum etc. It is divided in two parts- The Core genome: Set of genes common to all the genomes under study (These are often housekeeping genes vital for survival) and The Dispensable/Flexible Genome: Set of genes not present in all but one or some genomes under study. a bioinformatics tool BPGA can be used to characterize the Pan Genome of bacterial species.

Genetics of Disease

With the advent of next-generation sequencing we are obtaining enough sequence data to map the genes of complex diseases such as diabetes, infertility, breast cancer or Alzheimer's Disease. Genome-wide association studies are a useful approach to pinpoint the mutations responsible for such complex diseases. Through these studies, thousands of DNA variants have been identified that are associated with similar diseases and traits. Furthermore, the possibility for genes to be used at prognosis, diagnosis or treatment is one of the most essential applications. Many studies are discussing both the promising ways to choose the genes to be used and the problems and pitfalls of using genes to predict disease presence or prognosis.

Analysis of Mutations in Cancer

In cancer, the genomes of affected cells are rearranged in complex or even unpredictable ways. Massive sequencing efforts are used to identify previously unknown point mutations in a variety of genes in cancer. Bioinformaticians continue to produce specialized automated systems to manage the sheer volume of sequence data produced, and they create new algorithms and software to compare the sequencing results to the growing collection of human genome sequences and germline polymorphisms. New physical detection technologies are employed, such as oligonucleotide microarrays to identify chromosomal gains and losses (called comparative genomic hybridization), and single-nucleotide polymorphism arrays to detect known *point mutations*. These detection methods simultaneously measure several hundred thousand sites throughout the genome, and when used in high-throughput to measure thousands of samples, generate terabytes of data per experiment. Again the massive amounts and new types of data generate new opportunities for bioinformaticians. The data is often found to contain considerable variability, or noise, and thus Hidden Markov model and change-point analysis methods are being developed to infer real copy number changes.

With the breakthroughs that this next-generation sequencing technology is providing to the field of Bioinformatics, cancer genomics could drastically change. These new methods and software allow bioinformaticians to sequence many cancer genomes quickly and affordably. This could create a more flexible process for classifying types of cancer by analysis of cancer driven mutations in the genome. Furthermore, tracking of patients while the disease progresses may be possible in the future with the sequence of cancer samples.

Another type of data that requires novel informatics development is the analysis of lesions found to be recurrent among many tumors.

Gene and Protein Expression

Analysis of Gene Expression

The expression of many genes can be determined by measuring mRNA levels with multiple techniques including microarrays, expressed cDNA sequence tag (EST) sequencing, serial analysis of gene expression (SAGE) tag sequencing, massively parallel signature sequencing (MPSS), RNA-Seq, also known as "Whole Transcriptome Shotgun Sequencing" (WTSS), or various applications of multiplexed in-situ hybridization. All of these techniques are extremely noise-prone and/or subject to bias in the biological measurement, and a major research area in computational biology involves developing statistical tools to separate signal from noise in high-throughput gene expression studies. Such studies are often used to determine the genes implicated in a disorder: one might compare microarray data from cancerous epithelial cells to data from non-cancerous cells to determine the transcripts that are up-regulated and down-regulated in a particular population of cancer cells.

Analysis of Protein Expression

Protein microarrays and high throughput (HT) mass spectrometry (MS) can provide a snapshot of the proteins present in a biological sample. Bioinformatics is very much involved in making sense of protein microarray and HT MS data; the former approach faces similar problems as with microarrays targeted at mRNA, the latter involves the problem of matching large amounts of mass data against predicted masses from protein sequence databases, and the complicated statistical analysis of samples where multiple, but incomplete peptides from each protein are detected.

Analysis of Regulation

Regulation is the complex orchestration of events by which a signal, potentially an extracellular signal such as a hormone, eventually leads to an increase or decrease in the activity of one or more proteins. Bioinformatics techniques have been applied to explore various steps in this process.

For example, promoter analysis involves the identification and study of sequence motifs in the DNA surrounding the coding region of a gene. These motifs influence the extent to which that region is transcribed into mRNA.

Expression data can be used to infer gene regulation: one might compare microarray data from a wide variety of states of an organism to form hypotheses about the genes involved in each state. In a single-cell organism, one might compare stages of the cell cycle, along with various stress conditions (heat shock, starvation, etc.). One can then apply clustering algorithms to that expression data to determine which genes are co-expressed. For example, the upstream regions (promoters) of co-expressed genes can be searched for over-represented regulatory elements. Examples of clustering algorithms applied in gene clustering are k-means clustering, self-organizing maps (SOMs), hierarchical clustering, and consensus clustering methods.

Structural Bioinformatics

Protein structure prediction is another important application of bioinformatics. The amino acid sequence of a protein, the so-called primary structure, can be easily determined from the sequence on the gene that codes for it. In the vast majority of cases, this primary structure uniquely determines a structure in its native environment. (Of course, there are exceptions, such as the bovine spongiform encephalopathy – a.k.a. Mad Cow Disease – prion.) Knowledge of this structure is vital in understanding the function of the protein. Structural information is usually classified as one of *secondary*, *tertiary* and *quaternary* structure. A viable general solution to such predictions remains an open problem. Most efforts have so far been directed towards heuristics that work most of the time.

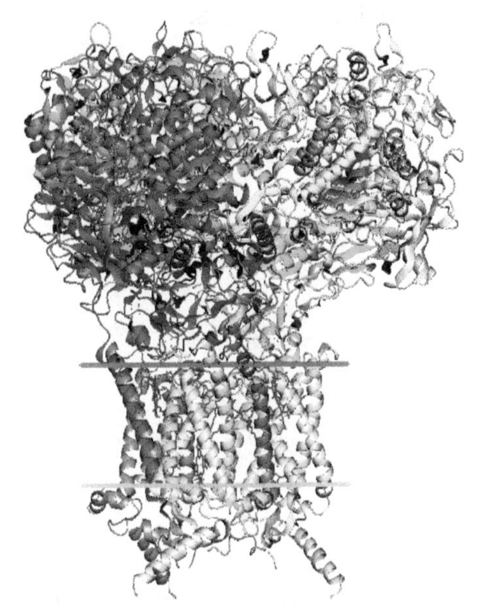

3-dimensional protein structures such as this one are common subjects in bioinformatic analyses.

One of the key ideas in bioinformatics is the notion of homology. In the genomic branch of bioinformatics, homology is used to predict the function of a gene: if the sequence of gene *A*, whose function is known, is homologous to the sequence of gene *B*, whose function is unknown, one could infer that B may share A's function. In the structural branch of bioinformatics, homology is used to determine which parts of a protein are important in structure formation and interaction with other proteins. In a technique called homology modeling, this information is used to predict the structure of a protein once the structure of a homologous protein is known. This currently remains the only way to predict protein structures reliably.

One example of this is the similar protein homology between hemoglobin in humans and the hemoglobin in legumes (leghemoglobin). Both serve the same purpose of transporting oxygen in the organism. Though both of these proteins have completely different amino acid sequences, their protein structures are virtually identical, which reflects their near identical purposes.

Other techniques for predicting protein structure include protein threading and *de novo* (from scratch) physics-based modeling.

Network and Systems Biology

Network analysis seeks to understand the relationships within biological networks such as metabolic or protein–protein interaction networks. Although biological networks can be constructed from a single type of molecule or entity (such as genes), network biology often attempts to integrate many different data types, such as proteins, small molecules, gene expression data, and others, which are all connected physically, functionally, or both.

Systems biology involves the use of computer simulations of cellular subsystems (such as the networks of metabolites and enzymes that comprise metabolism, signal transduction pathways and gene regulatory networks) to both analyze and visualize the complex connections of these cellular processes. Artificial life or virtual evolution attempts to understand evolutionary processes via the computer simulation of simple (artificial) life forms.

Molecular Interaction Networks

Interactions between proteins are frequently visualized and analyzed using networks. This network is made up of protein–protein interactions from *Treponema pallidum*, the causative agent of syphilis and other diseases.

Tens of thousands of three-dimensional protein structures have been determined by X-ray crystallography and protein nuclear magnetic resonance spectroscopy (protein NMR) and a central question in structural bioinformatics is whether it is practical to predict possible protein–protein interactions only based on these 3D shapes, without performing protein–protein interaction experiments. A variety of methods have been developed to tackle the protein–protein docking problem, though it seems that there is still much work to be done in this field.

Other interactions encountered in the field include Protein–ligand (including drug) and protein–peptide. Molecular dynamic simulation of movement of atoms about rotatable bonds is the fundamental principle behind computational algorithms, termed docking algorithms, for studying molecular interactions.

Others

Literature Analysis

The growth in the number of published literature makes it virtually impossible to read every paper, resulting in disjointed sub-fields of research. Literature analysis aims to employ computational and statistical linguistics to mine this growing library of text resources. For example:

- Abbreviation recognition – identify the long-form and abbreviation of biological terms
- Named entity recognition – recognizing biological terms such as gene names
- Protein–protein interaction – identify which proteins interact with which proteins from text

The area of research draws from statistics and computational linguistics.

High-throughput Image Analysis

Computational technologies are used to accelerate or fully automate the processing, quantification and analysis of large amounts of high-information-content biomedical imagery. Modern image analysis systems augment an observer's ability to make measurements from a large or complex set of images, by improving accuracy, objectivity, or speed. A fully developed analysis system may completely replace the observer. Although these systems are not unique to biomedical imagery, biomedical imaging is becoming more important for both diagnostics and research. Some examples are:

- high-throughput and high-fidelity quantification and sub-cellular localization (high-content screening, cytohistopathology, Bioimage informatics)
- morphometrics
- clinical image analysis and visualization
- determining the real-time air-flow patterns in breathing lungs of living animals
- quantifying occlusion size in real-time imagery from the development of and recovery during arterial injury
- making behavioral observations from extended video recordings of laboratory animals
- infrared measurements for metabolic activity determination
- inferring clone overlaps in DNA mapping, e.g. the Sulston score

High-throughput Single Cell Data Analysis

Computational techniques are used to analyse high-throughput, low-measurement single cell data, such as that obtained from flow cytometry. These methods typically involve finding populations of cells that are relevant to a particular disease state or experimental condition.

Biodiversity Informatics

Biodiversity informatics deals with the collection and analysis of biodiversity data, such as taxonomic databases, or microbiome data. Examples of such analyses include phylogenetics, niche modelling, species richness mapping, or species identification tools.

Databases

Databases are essential for bioinformatics research and applications. There is a huge number of available databases covering almost everything from DNA and protein sequences, molecular structures, to phenotypes and biodiversity. Databases generally fall into one of three types. Some contain data resulting directly from empirical methods such as gene knockouts. Others consist of predicted data, and most contain data from both sources. There are meta-databases that incorporate data compiled from multiple other databases. Some others are specialized, such as those specific to an organism. These databases vary in their format, way of accession and whether they are public or not. Some of the most commonly used databases are listed below. For a more comprehensive list, please check the link at the beginning of the subsection.

- Used in Motif Finding: GenomeNet MOTIF Search
- Used in Gene Ontology: ToppGene FuncAssociate, Enrichr, GATHER
- Used in Gene Finding: Hidden Markov Model
- Used in finding Protein Structures/Family: PFAM
- Used for Next Generation Sequencing: (Not database but data format), FASTQ Format
- Used in Gene Expression Analysis: GEO, ArrayExpress
- Used in Network Analysis: Interaction Analysis Databases(BioGRID, MINT, HPRD, Curated Human Signaling Network), Functional Networks (STRING, KEGG)
- Used in design of synthetic genetic circuits: GenoCAD

Please keep in mind that this is a quick sampling and generally most computation data is supported by wet lab data as well.

Software and Tools

Software tools for bioinformatics range from simple command-line tools, to more complex graphical programs and standalone web-services available from various bioinformatics companies or public institutions.

Open-source Bioinformatics Software

Many free and open-source software tools have existed and continued to grow since the 1980s. The combination of a continued need for new algorithms for the analysis of emerging types of biological readouts, the potential for innovative *in silico* experiments, and freely available open code bases have helped to create opportunities for all research groups to contribute to both bioinformatics and the range of open-source software available, regardless of their funding arrangements. The open source tools often act as incubators of ideas, or community-supported plug-ins in commercial applications. They may also provide *de facto* standards and shared object models for assisting with the challenge of bioinformation integration.

The range of open-source software packages includes titles such as Bioconductor, BioPerl, Biopython, BioJava, BioJS, BioRuby, Bioclipse, EMBOSS, .NET Bio, Orange with its bioinformatics add-on, Apache Taverna, UGENE and GenoCAD. To maintain this tradition and create further opportunities, the non-profit Open Bioinformatics Foundation have supported the annual Bioinformatics Open Source Conference (BOSC) since 2000.

An alternative method to build public bioinformatics databases is to use the MediaWiki engine with the *WikiOpener* extension. This system allows the database to be accessed and updated by all experts in the field.

Web Services in Bioinformatics

SOAP- and REST-based interfaces have been developed for a wide variety of bioinformatics applications allowing an application running on one computer in one part of the world to use algorithms, data and computing resources on servers in other parts of the world. The main advantages derive from the fact that end users do not have to deal with software and database maintenance overheads.

Basic bioinformatics services are classified by the EBI into three categories: SSS (Sequence Search Services), MSA (Multiple Sequence Alignment), and BSA (Biological Sequence Analysis). The availability of these service-oriented bioinformatics resources demonstrate the applicability of web-based bioinformatics solutions, and range from a collection of standalone tools with a common data format under a single, standalone or web-based interface, to integrative, distributed and extensible bioinformatics workflow management systems.

Bioinformatics Workflow Management Systems

A Bioinformatics workflow management system is a specialized form of a workflow management system designed specifically to compose and execute a series of computational or data manipulation steps, or a workflow, in a Bioinformatics application. Such systems are designed to

- provide an easy-to-use environment for individual application scientists themselves to create their own workflows
- provide interactive tools for the scientists enabling them to execute their workflows and view their results in real-time
- simplify the process of sharing and reusing workflows between the scientists.
- enable scientists to track the provenance of the workflow execution results and the workflow creation steps.

Some of the platforms giving this service: Galaxy, Kepler, Taverna, UGENE, Anduril.

Education Platforms

Software platforms designed to teach bioinformatics concepts and methods include Rosalind and online courses offered through the Swiss Institute of Bioinformatics Training Portal. The Canadian Bioinformatics Workshops provides videos and slides from training workshops on their website under a Creative Commons license. The 4273π project or 4273pi project also offers open source educational materials for free. The course runs on low cost raspberry pi computers and has been used to teach adults and school pupils. 4273π is actively developed by a consortium of academics and research staff who have run research level bioinformatics using raspberry pi computers and the 4273π operating system.

MOOC platforms also provide online certifications in bioinformatics and related disciplines, including Coursera's Bioinformatics Specialization (UC San Diego) and Genomic Data Science Specialization (Johns Hopkins) as well as EdX's Data Analysis for Life Sciences XSeries (Harvard).

Conferences

There are several large conferences that are concerned with bioinformatics. Some of the most notable examples are Intelligent Systems for Molecular Biology (ISMB), European Conference on Computational Biology (ECCB), and Research in Computational Molecular Biology (RECOMB).

Central Dogma of Molecular Biology

The central dogma of molecular biology is an explanation of the flow of genetic information within a biological system. It was first stated by Francis Crick in 1956 and re-stated in a *Nature* paper published in 1970:

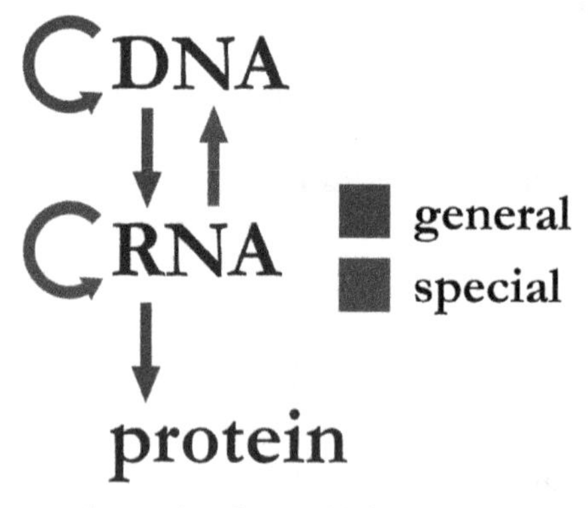

Information flow in biological systems

"The central dogma of molecular biology deals with the detailed residue-by-residue transfer of sequential information. It states that such information cannot be transferred back from protein to either protein or nucleic acid."

—Francis Crick

The central dogma has also been described as "DNA makes RNA and RNA makes protein," a positive statement which was originally termed the sequence hypothesis by Crick. However, this simplification does not make it clear that the central dogma as stated by Crick does not preclude the reverse flow of information from RNA to DNA, only ruling out the flow from protein to RNA or DNA. Crick's use of the word dogma was unconventional, and has been controversial.

The dogma is a framework for understanding the transfer of sequence information between information-carrying biopolymers, in the most common or general case, in living organisms. There are 3 major classes of such biopolymers: DNA and RNA (both nucleic acids), and protein. There are 3×3 = 9 conceivable direct transfers of information that can occur between these. The dogma classes these into 3 groups of 3: 3 general transfers (believed to occur normally in most cells), 3 special transfers (known to occur, but only under specific conditions in case of some viruses or in a laboratory), and 3 unknown transfers (believed never to occur). The general transfers describe the normal flow of biological information: DNA can be copied to DNA (DNA replication), DNA information can be copied into mRNA (transcription), and proteins can be synthesized using the information in mRNA as a template (translation).

Biological Sequence Information

The biopolymers that comprise DNA, FNA, RNA and (poly)peptides are linear polymers (i.e.: each monomer is connected to at most two other monomers). The sequence of their monomers effectively encodes information. The transfers of information described by the central dogma ideally are faithful, deterministic transfers, wherein one biopolymer's sequence is used as a template for the construction of another biopolymer with a sequence that is entirely dependent on the original biopolymer's sequence.

General Transfers of Biological Sequential Information

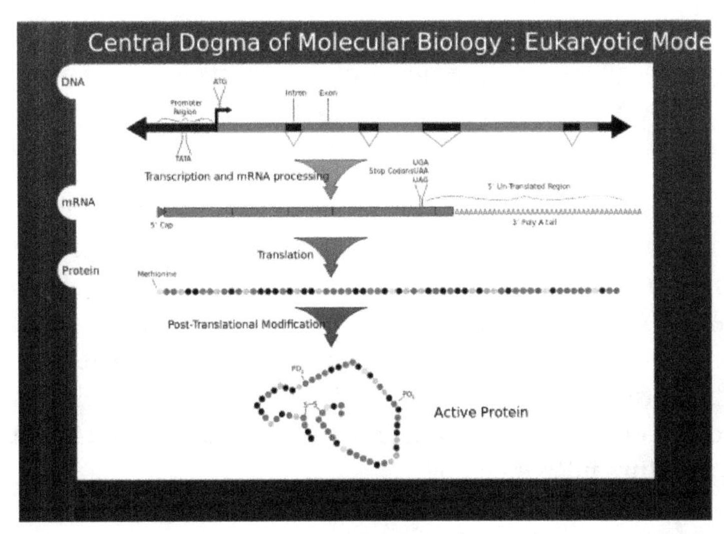

Table of the 3 classes of information transfer suggested by the dogma		
General	**Special**	**Unknown**
DNA → DNA	RNA → DNA	protein → DNA
DNA → RNA	RNA → RNA	protein → RNA
RNA → protein	DNA → protein	protein → protein

DNA Replications

In the sense that DNA replication must occur if genetic material is to be provided for the progeny of any cell, whether somatic or reproductive, the copying from DNA to DNA arguably is the fundamental step in the central dogma. A complex group of proteins called the replisome performs the replication of the information from the parent strand to the complementary daughter strand.

The replisome comprises:

- a helicase that unwinds the superhelix as well as the double-stranded DNA helix to create a replication fork

- SSB protein that binds open the double-stranded DNA to prevent it from reassociating
- RNA primase that adds a complementary RNA primer to each template strand as a starting point for replication
- DNA polymerase III that reads the existing template chain from its 3' end to its 5' end and adds new complementary nucleotides from the 5' end to the 3' end of the daughter chain
- DNA polymerase I that removes the RNA primers and replaces them with DNA.
- DNA ligase that joins the two Okazaki fragments with phosphodiester bonds to produce a continuous chain.

This process typically takes place during S phase of the cell cycle.

Transcription

Transcription is the process by which the information contained in a section of DNA is replicated in the form of a newly assembled piece of messenger RNA (mRNA). Enzymes facilitating the process include RNA polymerase and transcription factors. In eukaryotic cells the primary transcript is (pre-mRNA). Pre-mRNA must be processed for translation to proceed. Processing includes the addition of a 5' cap and a poly-A tail to the pre-mRNA chain, followed by splicing. Alternative splicing occurs when appropriate, increasing the diversity of the proteins that any single mRNA can produce. The product of the entire transcription process that began with the production of the pre-mRNA chain, is a mature mRNA chain.

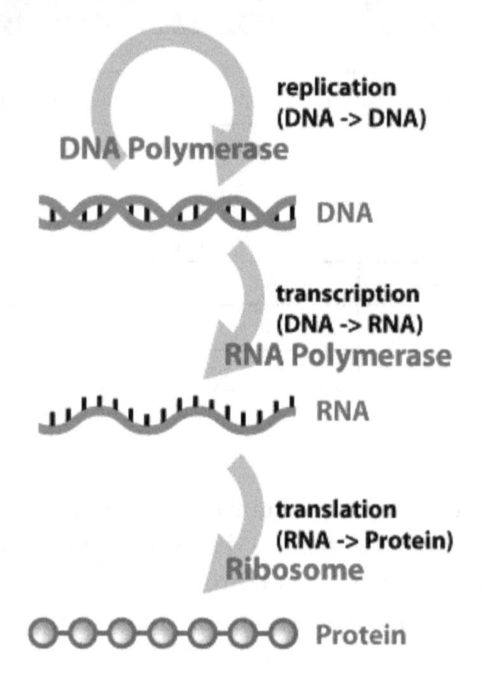

Translation

The mature mRNA finds its way to a ribosome, where it gets translated. In prokaryotic cells, which have no nuclear compartment, the processes of transcription and translation may be linked together without clear separation. In eukaryotic cells, the site of transcription (the cell nucleus) is usually separated from the site of translation (the cytoplasm), so the mRNA must be transported out of the nucleus into the cytoplasm, where it can be bound by ribosomes. The ribosome reads the mRNA triplet codons, usually beginning with an AUG (adenine–uracil–guanine), or initiator methionine codon downstream of the ribosome binding site. Complexes of initiation factors and elongation factors bring aminoacylated transfer RNAs (tRNAs) into the ribosome-mRNA complex, matching the codon in the mRNA to the anti-codon on the tRNA. Each tRNA bears the appropriate amino acid residue to add to the polypeptide chain being synthesised. As the amino acids get linked into the growing peptide chain, the chain begins folding into the correct conformation. Translation ends with a stop codon which may be a UAA, UGA, or UAG triplet.

The mRNA does not contain all the information for specifying the nature of the mature protein. The nascent polypeptide chain released from the ribosome commonly requires additional processing before the final product emerges. For one thing, the correct folding process is complex and vitally important. For most proteins it requires other chaperone proteins to control the form of the product. Some proteins then excise internal segments from their own peptide chains, splicing the free ends that border the gap; in such processes the inside "discarded" sections are called inteins. Other proteins must be split into multiple sections without splicing. Some polypeptide chains need to be cross-linked, and others must be attached to cofactors such as haem (heme) before they become functional.

Special Transfers of Biological Sequential Information

Reverse Transcription

Reverse transcription is the transfer of information from RNA to DNA (the reverse of normal transcription). This is known to occur in the case of retroviruses, such as HIV, as well as in eukaryotes, in the case of retrotransposons and telomere synthesis. It is the process by which genetic information from RNA gets transcribed into new DNA.

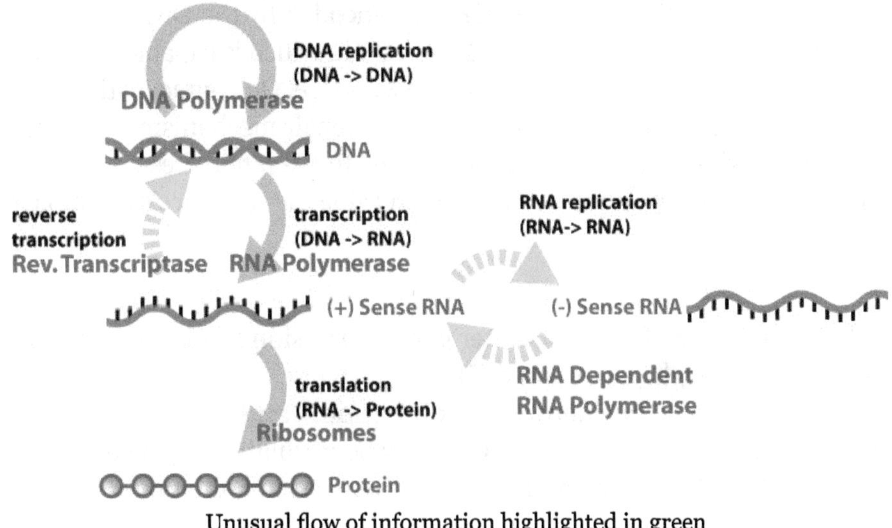

Unusual flow of information highlighted in green

RNA Replication

RNA replication is the copying of one RNA to another. Many viruses replicate this way. The enzymes that copy RNA to new RNA, called RNA-dependent RNA polymerases, are also found in many eukaryotes where they are involved in RNA silencing.

RNA editing, in which an RNA sequence is altered by a complex of proteins and a "guide RNA", could also be seen as an RNA-to-RNA transfer.

Direct Translation from DNA to Protein

Direct translation from DNA to protein has been demonstrated in a cell-free system (i.e. in a test tube), using extracts from *E. coli* that contained ribosomes, but not intact cells. These cell fragments could synthesize proteins from single-stranded DNA templates isolated from other organisms (e,g., mouse or toad), and neomycin was found to enhance this effect. However, it was unclear whether this mechanism of translation corresponded specifically to the genetic code.

Transfers of Information not Explicitly Covered in the Theory

Posttranslational Modification

After protein amino acid sequences have been translated from nucleic acid chains, they can be edited by appropriate enzymes. Although this is a form of protein affecting protein sequence, not explicitly covered by the central dogma, there are not many clear examples where the associated concepts of the two fields have much to do with each other.

Inteins

An intein is a "parasitic" segment of a protein that is able to excise itself from the chain of amino acids as they emerge from the ribosome and rejoin the remaining portions with a peptide bond in such a manner that the main protein "backbone" does not fall apart. This is a case of a protein changing its own primary sequence from the sequence originally encoded by the DNA of a gene. Additionally, most inteins contain a homing endonuclease or HEG domain which is capable of finding a copy of the parent gene that does not include the intein nucleotide sequence. On contact with the intein-free copy, the HEG domain initiates the DNA double-stranded break repair mechanism. This process causes the intein sequence to be copied from the original source gene to the intein-free gene. This is an example of protein directly editing DNA sequence, as well as increasing the sequence's heritable propagation.

Methylation

Variation in methylation states of DNA can alter gene expression levels significantly. Methylation variation usually occurs through the action of DNA methylases. When the change is heritable, it is considered epigenetic. When the change in information status is not heritable, it would be a somatic epitype. The effective information content has been changed by means of the actions of a protein or proteins on DNA, but the primary DNA sequence is not altered.

Prions

Prions are proteins of particular amino acid sequences in particular conformations. They propagate themselves in host cells by making conformational changes in other molecules of protein with the same amino acid sequence, but with a different conformation that is functionally important to the cell. Once the protein has been transconformed to the prion folding it changes function. In turn it can convey information into new cells and reconfigure more functional molecules of that sequence into the alternate prion form. In some types of prion in fungi this change is continuous and direct; the information flow is Protein → Protein.

Some scientists such as Alain E. Bussard and Eugene Koonin have argued that prion-mediated inheritance violates the central dogma of molecular biology. However, Rosalind Ridley in *Molecular Pathology of the Prions* (2001) has written that "The prion hypothesis is not heretical to the central dogma of molecular biology—that the information necessary to manufacture proteins is encoded in the nucleotide sequence of nucleic acid—because it does not claim that proteins replicate. Rather, it claims that there is a source of information within protein molecules that contributes to their biological function, and that this information can be passed on to other molecules."

Natural Genetic Engineering

James A. Shapiro argues that a superset of these examples should be classified as natural genetic engineering and are sufficient to falsify the central dogma. While Shapiro has received a respectful hearing for his view, his critics have not been convinced that his reading of the central dogma is in line with what Crick intended.

Use of the Term "Dogma"

In his autobiography, *What Mad Pursuit*, Crick wrote about his choice of the word *dogma* and some of the problems it caused him:

"I called this idea the central dogma, for two reasons, I suspect. I had already used the obvious word hypothesis in the sequence hypothesis, and in addition I wanted to suggest that this new assumption was more central and more powerful. ... As it turned out, the use of the word dogma caused almost more trouble than it was worth. Many years later Jacques Monod pointed out to me that I did not appear to understand the correct use of the word dogma, which is a belief *that cannot be doubted*. I did apprehend this in a vague sort of way but since I thought that *all* religious beliefs were without foundation, I used the word the way I myself thought about it, not as most of the world does, and simply applied it to a grand hypothesis that, however plausible, had little direct experimental support."

Similarly, Horace Freeland Judson records in *The Eighth Day of Creation*:

"My mind was, that a dogma was an idea for which there was *no reasonable evidence*. You see?!" And Crick gave a roar of delight. "I just didn't *know* what dogma *meant*. And I could just as well have called it the 'Central Hypothesis,' or — you know. Which is what I meant to say. Dogma was just a catch phrase."

References

- Wong, KC (2016). Computational Biology and Bioinformatics: Gene Regulation. CRC Press (Taylor & Francis Group). ISBN 9781498724975.

- Moody, Glyn (2004). Digital Code of Life: How Bioinformatics is Revolutionizing Science, Medicine, and Business. ISBN 978-0-471-32788-2.
- Hye-Jung, E.C.; Jaswinder, K.; Martin, K.; Samuel, A.A; Marco, A.M (2014). “”Second-Generation Sequencing for Cancer Genome Analysis””. In Dellaire, Graham; Berman, Jason N.; Arceci, Robert J. Cancer Genomics. Boston (US): Academic Press. pp. 13–30. doi:10.1016/B978-0-12-396967-5.00002-5. ISBN 9780123969675.
- Horace Freeland Judson (1996). “Chapter 6: My mind was, that a dogma was an idea for which there was no reasonable evidence. You see?!”. The Eighth Day of Creation: Makers of the Revolution in Biology (25th anniversary edition). Cold Spring Harbor, NY: Cold Spring Harbor Laboratory Press. ISBN 0-87969-477-7
- Brohée, Sylvain; Barriot, Roland; Moreau, Yves. “Biological knowledge bases using Wikis: combining the flexibility of Wikis with the structure of databases”. Bioinformatics. Oxford Journals. Retrieved 5 May 2015.
- Nisbet, Robert (14 May 2009). “BIOINFORMATICS”. Handbook of Statistical Analysis and Data Mining Applications. John Elder IV, Gary Miner. Academic Press. p. 328. Retrieved 9 May 2014.
- Attwood TK, Gisel A, Eriksson NE, Bongcam-Rudloff E (2011). “Concepts, Historical Milestones and the Central Place of Bioinformatics in Modern Biology: A European Perspective”. Bioinformatics – Trends and Methodologies. InTech. Retrieved 8 Jan 2012.

6

Macromolecule Blotting: A Comprehensive Study

Blotting can be divided into three techniques, southern blotting, northern blotting and western blotting. The macromolecule analyzed in southern blot technique is DNA, for northern blot it is the RNA and for western it is protein. The chapter on macromolecule blotting offers an insightful focus, keeping in mind the complex subject matter.

Southern Blot

A Southern blot is a method used in molecular biology for detection of a specific DNA sequence in DNA samples. Southern blotting combines transfer of electrophoresis-separated DNA fragments to a filter membrane and subsequent fragment detection by probe hybridization.

The method is named after its inventor, the British biologist Edwin Southern. Other blotting methods (i.e., western blot, northern blot, eastern blot, southwestern blot) that employ similar principles, but using RNA or protein, have later been named in reference to Edwin Southern's name. As the label is eponymous, Southern is capitalised, as is conventional for proper nouns. The names for other blotting methods may follow this convention, by analogy.

Method

1. Restriction endonucleases are used to cut high-molecular-weight DNA strands into smaller fragments.
2. The DNA fragments are then electrophoresed on an agarose gel to separate them by size.
3. If some of the DNA fragments are larger than 15 kb, then prior to blotting, the gel may be treated with an acid, such as dilute HCl. This depurinates the DNA fragments, breaking the DNA into smaller pieces, thereby allowing more efficient transfer from the gel to membrane.
4. If alkaline transfer methods are used, the DNA gel is placed into an alkaline solution (typically containing sodium hydroxide) to denature the double-stranded DNA. The denaturation in an alkaline environment may improve binding of the negatively charged thymine residues of DNA to a positively charged amino groups of membrane, separating it into single DNA strands for later hybridization to the probe, and destroys any residual RNA that may still be present in the DNA. The choice of alkaline over neutral transfer methods, however, is often empirical and may result in equivalent results.

5. A sheet of nitrocellulose (or, alternatively, nylon) membrane is placed on top of (or below, depending on the direction of the transfer) the gel. Pressure is applied evenly to the gel (either using suction, or by placing a stack of paper towels and a weight on top of the membrane and gel), to ensure good and even contact between gel and membrane. If transferring by suction, 20X SSC buffer is used to ensure a seal and prevent drying of the gel. Buffer transfer by capillary action from a region of high water potential to a region of low water potential (usually filter paper and paper tissues) is then used to move the DNA from the gel onto the membrane; ion exchange interactions bind the DNA to the membrane due to the negative charge of the DNA and positive charge of the membrane.

6. The membrane is then baked in a vacuum or regular oven at 80 °C for 2 hours (standard conditions; nitrocellulose or nylon membrane) or exposed to ultraviolet radiation (nylon membrane) to permanently attach the transferred DNA to the membrane.

7. The membrane is then exposed to a hybridization probe—a single DNA fragment with a specific sequence whose presence in the target DNA is to be determined. The probe DNA is labelled so that it can be detected, usually by incorporating radioactivity or tagging the molecule with a fluorescent or chromogenic dye. In some cases, the hybridization probe may be made from RNA, rather than DNA. To ensure the specificity of the binding of the probe to the sample DNA, most common hybridization methods use salmon or herring sperm DNA for blocking of the membrane surface and target DNA, deionized formamide, and detergents such as SDS to reduce non-specific binding of the probe.

8. After hybridization, excess probe is washed from the membrane (typically using SSC buffer), and the pattern of hybridization is visualized on X-ray film by autoradiography in the case of a radioactive or fluorescent probe, or by development of colour on the membrane if a chromogenic detection method is used.

Result

Hybridization of the probe to a specific DNA fragment on the filter membrane indicates that this fragment contains DNA sequence that is complementary to the probe. The transfer step of the DNA from the electrophoresis gel to a membrane permits easy binding of the labeled hybridization probe to the size-fractionated DNA. It also allows for the fixation of the target-probe hybrids, required for analysis by autoradiography or other detection methods. Southern blots performed with restriction enzyme-digested genomic DNA may be used to determine the number of sequences (e.g., gene copies) in a genome. A probe that hybridizes only to a single DNA segment that has not been cut by the restriction enzyme will produce a single band on a Southern blot, whereas multiple bands will likely be observed when the probe hybridizes to several highly similar sequences (e.g., those that may be the result of sequence duplication). Modification of the hybridization conditions (for example, increasing the hybridization temperature or decreasing salt concentration) may be used to increase specificity and decrease hybridization of the probe to sequences that are less than 100% similar.

Applications

Southern transfer may be used for homology-based cloning on the basis of amino acid sequence of the protein product of the target gene. Oligonucleotides are designed that are similar to the target sequence. The oligonucleotides are chemically synthesised, radiolabeled, and used to screen a DNA library, or other collections of cloned DNA fragments. Sequences that hybridize with the hybridization probe are further analysed, for example, to obtain the full length sequence of the targeted gene.

Southern blotting can also be used to identify methylated sites in particular genes. Particularly useful are the restriction nucleases *MspI* and *HpaII*, both of which recognize and cleave within the same sequence. However, *HpaII* requires that a C within that site be methylated, whereas *MspI* cleaves only DNA unmethylated at that site. Therefore, any methylated sites within a sequence analyzed with a particular probe will be cleaved by the former, but not the latter, enzyme.

Northern Blot

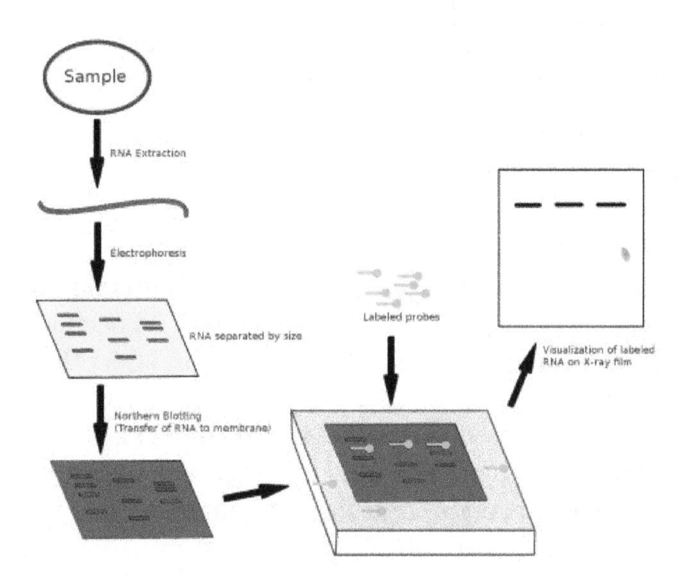

Flow diagram outlining the general procedure for RNA detection by northern blotting.

The northern blot is a technique used in molecular biology research to study gene expression by detection of RNA (or isolated mRNA) in a sample.

With northern blotting it is possible to observe cellular control over structure and function by determining the particular gene expression rates during differentiation and morphogenesis, as well as in abnormal or diseased conditions. Northern blotting involves the use of electrophoresis to separate RNA samples by size, and detection with a hybridization probe complementary to part of or the entire target sequence. The term 'northern blot' actually refers specifically to the capillary transfer of RNA from the electrophoresis gel to the blotting membrane. However, the entire process is commonly referred to as northern blotting. The northern blot technique was developed in 1977 by James Alwine, David Kemp, and George Stark at Stanford University. Northern blotting takes its name from its similarity to the first blotting technique, the Southern blot, named for biologist Edwin Southern. The major difference is that RNA, rather than DNA, is analyzed in the northern blot.

Procedure

A general blotting procedure starts with extraction of total RNA from a homogenized tissue sample or from cells. Eukaryotic mRNA can then be isolated through the use of oligo (dT) cellulose chromatography to isolate only those RNAs with a poly(A) tail. RNA samples are then separated by gel electrophoresis. Since the gels are fragile and the probes are unable to enter the matrix, the RNA samples, now separated by size, are transferred to a nylon membrane through a capillary or vacuum blotting system.

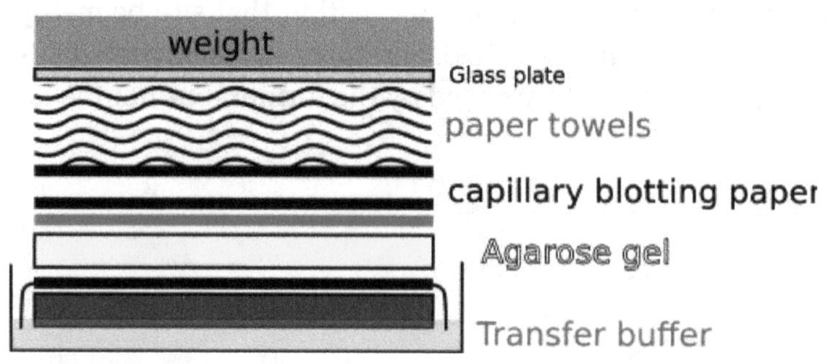

Capillary blotting system setup for the transfer of RNA from an electrophoresis gel to a blotting membrane.

A nylon membrane with a positive charge is the most effective for use in northern blotting since the negatively charged nucleic acids have a high affinity for them. The transfer buffer used for the blotting usually contains formamide because it lowers the annealing temperature of the probe-RNA interaction, thus eliminating the need for high temperatures, which could cause RNA degradation. Once the RNA has been transferred to the membrane, it is immobilized through covalent linkage to the membrane by UV light or heat. After a probe has been labeled, it is hybridized to the RNA on the membrane. Experimental conditions that can affect the efficiency and specificity of hybridization include ionic strength, viscosity, duplex length, mismatched base pairs, and base composition. The membrane is washed to ensure that the probe has bound specifically and to prevent background signals from arising. The hybrid signals are then detected by X-ray film and can be quantified by densitometry. To create controls for comparison in a northern blot, samples not displaying the gene product of interest can be used after determination by microarrays or RT-PCR.

Gels

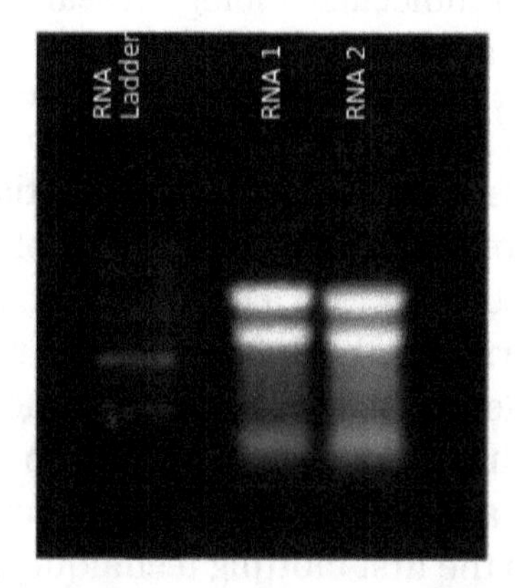

RNA run on a formaldehyde agarose gel to highlight the 28S (top band) and 18S (lower band) ribosomal subunits.

The RNA samples are most commonly separated on agarose gels containing formaldehyde as a denaturing agent for the RNA to limit secondary structure. The gels can be stained with ethidium bromide (EtBr) and viewed under UV light to observe the quality and quantity of RNA before blotting. Polyacrylamide gel electrophoeresis with urea can also be used in RNA separation but it is most commonly used for fragmented RNA or microRNAs. An RNA ladder is often run alongside the samples on an electrophoresis gel to observe the size of fragments obtained but in total RNA samples the ribosomal subunits can act as size markers. Since the large ribosomal subunit is 28S (approximately 5kb) and the small ribosomal subunit is 18S (approximately 2kb) two prominent bands appear on the gel, the larger at close to twice the intensity of the smaller.

Probes

Probes for northern blotting are composed of nucleic acids with a complementary sequence to all or part of the RNA of interest, they can be DNA, RNA, or oligonucleotides with a minimum of 25 complementary bases to the target sequence. RNA probes (riboprobes) that are transcribed in vitro are able to withstand more rigorous washing steps preventing some of the background noise. Commonly cDNA is created with labelled primers for the RNA sequence of interest to act as the probe in the northern blot. The probes must be labelled either with radioactive isotopes (^{32}P) or with chemiluminescence in which alkaline phosphatase or horseradish peroxidase (HRP) break down chemiluminescent substrates producing a detectable emission of light. The chemiluminescent labelling can occur in two ways: either the probe is attached to the enzyme, or the probe is labelled with a ligand (e.g. biotin) for which the ligand (e.g., avidin or streptavidin) is attached to the enzyme (e.g. HRP). X-ray film can detect both the radioactive and chemiluminescent signals and many researchers prefer the chemiluminescent signals because they are faster, more sensitive, and reduce the health hazards that go along with radioactive labels. The same membrane can be probed up to five times without a significant loss of the target RNA.

Applications

Northern blotting allows one to observe a particular gene's expression pattern between tissues, organs, developmental stages, environmental stress levels, pathogen infection, and over the course of treatment. The technique has been used to show overexpression of oncogenes and downregulation of tumor-suppressor genes in cancerous cells when compared to 'normal' tissue, as well as the gene expression in the rejection of transplanted organs. If an upregulated gene is observed by an abundance of mRNA on the northern blot the sample can then be sequenced to determine if the gene is known to researchers or if it is a novel finding. The expression patterns obtained under given conditions can provide insight into the function of that gene. Since the RNA is first separated by size, if only one probe type is used variance in the level of each band on the membrane can provide insight into the size of the product, suggesting alternative splice products of the same gene or repetitive sequence motifs. The variance in size of a gene product can also indicate deletions or errors in transcript processing. By altering the probe target used along the known sequence it is possible to determine which region of the RNA is missing.

BlotBase is an online database publishing northern blots. BlotBase has over 700 published northern blots of human and mouse samples, in over 650 genes across more than 25 different tissue types. Northern blots can be searched by a blot ID, paper reference, gene identifier, or by tissue. The results of a search provide the blot ID, species, tissue, gene, expression level, blot image (if available), and links to the publication that the work originated from. This new database provides sharing of information between members of the science community that was not previously seen in northern blotting as it was in sequence analysis, genome determination, protein structure, etc.

Advantages and Disadvantages

Analysis of gene expression can be done by several different methods including RT-PCR, RNase protection assays, microarrays, RNA-Seq, serial analysis of gene expression (SAGE), as well as northern blotting. Microarrays are quite commonly used and are usually consistent with data obtained from northern blots; however, at times northern blotting is able to detect small changes in gene expression that microarrays cannot. The advantage that microarrays have over northern blots is that thousands of genes can be visualized at a time, while northern blotting is usually looking at one or a small number of genes.

A problem in northern blotting is often sample degradation by RNases (both endogenous to the sample and through environmental contamination), which can be avoided by proper sterilization of glassware and the use of RNase inhibitors such as DEPC (diethylpyrocarbonate). The chemicals used in most northern blots can be a risk to the researcher, since formaldehyde, radioactive material, ethidium bromide, DEPC, and UV light are all harmful under certain exposures. Compared to RT-PCR, northern blotting has a low sensitivity, but it also has a high specificity, which is important to reduce false positive results.

The advantages of using northern blotting include the detection of RNA size, the observation of alternate splice products, the use of probes with partial homology, the quality and quantity of RNA can be measured on the gel prior to blotting, and the membranes can be stored and reprobed for years after blotting.

For northern blotting for the detection of acetylcholinesterase mRNA the nonradioactive technique was compared to a radioactive technique and found as sensitive as the radioactive one, but requires no protection against radiation and is less time consuming.

Reverse Northern Blot

Researchers occasionally use a variant of the procedure known as the reverse northern blot. In this procedure, the substrate nucleic acid (that is affixed to the membrane) is a collection of isolated DNA fragments, and the probe is RNA extracted from a tissue and radioactively labelled.

The use of DNA microarrays that have come into widespread use in the late 1990s and early 2000s is more akin to the reverse procedure, in that they involve the use of isolated DNA fragments affixed to a substrate, and hybridization with a probe made from cellular RNA. Thus the reverse procedure, though originally uncommon, enabled northern analysis to evolve into gene expression profiling, in which many (possibly all) of the genes in an organism may have their expression monitored.

Western Blot

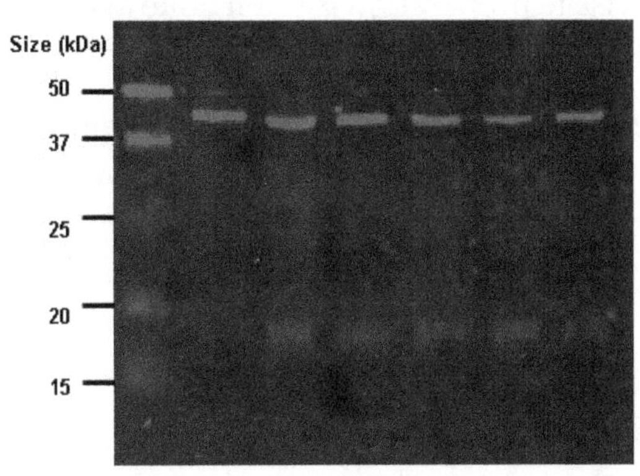

Western blot using an antibody that recognizes proteins modified with lipoic acid.

The western blot (sometimes called the protein immunoblot) is a widely used analytical technique used in molecular biology to detect specific proteins in a sample of tissue homogenate or extract. It uses gel electrophoresis to separate native proteins by 3-D structure or denatured proteins by the length of the polypeptide. The proteins are then transferred to a membrane (typically nitrocellulose or PVDF), where they are stained with antibodies specific to the target protein. The gel electrophoresis step is included in western blot analysis to resolve the issue of the cross-reactivity of antibodies.

There are many reagent companies that specialize in providing antibodies (both monoclonal and polyclonal antibodies) against tens of thousands of different proteins. Commercial antibodies can be expensive, although the unbound antibody can be reused between experiments. This method is used in the fields of molecular biology, immunogenetics and other molecular biology disciplines. A number of search engines, such as CiteAb, Antibodypedia, and SeekProducts, are available that can help researchers find suitable antibodies for use in western blotting.

Other related techniques include dot blot analysis, immunohistochemistry and immunocytochemistry where antibodies are used to detect proteins in tissues and cells by immunostaining, and enzyme-linked immunosorbent assay (ELISA).

The method originated in the laboratory of Harry Towbin at the Friedrich Miescher Institute. The name *western blot* was given to the technique by W. Neal Burnette and is a play on the name Southern blot, a technique for DNA detection developed earlier by Edwin Southern. Detection of RNA is termed northern blot and was developed by George Stark at Stanford.

Steps

Tissue Preparation

Samples can be taken from whole tissue or from cell culture. Solid tissues are first broken down mechanically using a blender (for larger sample volumes), using a homogenizer (smaller volumes), or by sonication. Cells may also be broken open by one of the above mechanical methods. However, virus or environmental samples can be the source of protein and thus western blotting is not restricted to cellular studies only.

Assorted detergents, salts, and buffers may be employed to encourage lysis of cells and to solubilize proteins. Protease and phosphatase inhibitors are often added to prevent the digestion of the sample by its own enzymes. Tissue preparation is often done at cold temperatures to avoid protein denaturing and degradation.

A combination of biochemical and mechanical techniques – comprising various types of filtration and centrifugation – can be used to separate different cell compartments and organelles.

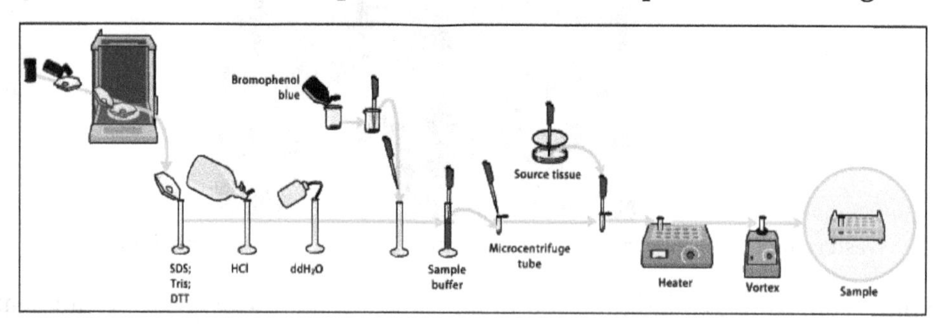

Gel Electrophoresis

The proteins of the sample are separated using gel electrophoresis. Separation of proteins may be by isoelectric point (pI), molecular weight, electric charge, or a combination of these factors. The nature of the separation depends on the treatment of the sample and the nature of the gel. This is a very useful way to identify a protein.

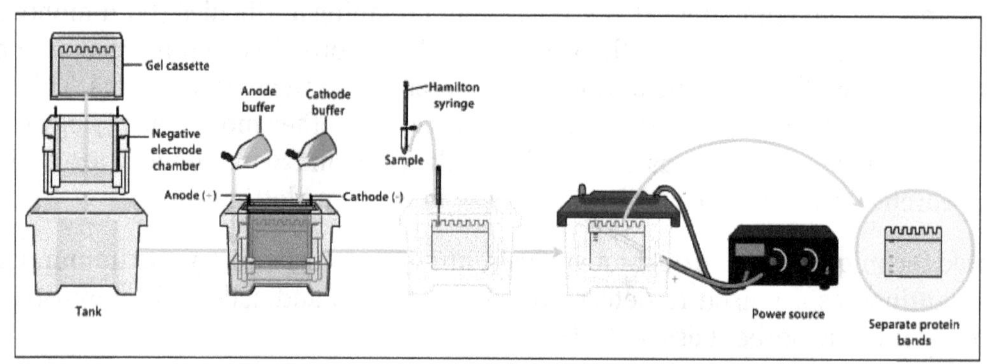

It is also possible to use a two-dimensional (2-D) gel which spreads the proteins from a single sample out in two dimensions. Proteins are separated according to isoelectric point (pH at which they have neutral net charge) in the first dimension, and according to their molecular weight in the second dimension.

By far the most common type of gel electrophoresis employs polyacrylamide gels and buffers loaded with sodium dodecyl sulfate (SDS). SDS-PAGE (SDS polyacrylamide gel electrophoresis) maintains polypeptides in a denatured state once they have been treated with strong reducing agents to remove secondary and tertiary structure (e.g. disulfide bonds [S-S] to sulfhydryl groups [SH and SH]) and thus allows separation of proteins by their molecular weight. Sampled proteins become covered in the negatively charged SDS and move to the positively charged electrode through the acrylamide mesh of the gel. Smaller proteins migrate faster through this mesh and the proteins are thus separated according to size (usually measured in kilodaltons, kDa). The concentration of acrylamide determines the resolution of the gel - the greater the acrylamide concentration, the better the resolution of lower molecular weight proteins. The lower the acrylamide concentration, the better the resolution of higher molecular weight proteins. Proteins travel only in one dimension along the gel for most blots.

Samples are loaded into *wells* in the gel. One lane is usually reserved for a *marker* or *ladder*, a commercially available mixture of proteins having defined molecular weights, typically stained so as to form visible, coloured bands. When voltage is applied along the gel, proteins migrate through it at different speeds dependent on their size. These different rates of advancement (different *electrophoretic mobilities*) separate into *bands* within each *lane*.

Transfer

In order to make the proteins accessible to antibody detection, they are moved from within the gel onto a membrane made of *nitrocellulose or polyvinylidene difluoride (PVDF)*. The primary method for transferring the proteins is called electroblotting and uses an electric current to pull proteins from the gel into the PVDF or nitrocellulose membrane. The proteins move from within the gel onto the membrane while maintaining the organization they had within the gel. An older method of transfer involves placing a membrane on top of the gel, and a stack of filter papers on top of that. The entire stack is placed in a buffer solution which moves up the paper by capillary action, bringing the proteins with it. In practice this method is not used as it takes too much time; electroblotting is preferred. As a result of either "blotting" process, the proteins are exposed on a thin surface layer for detection. Both varieties of membrane are chosen for their non-specific protein binding properties (i.e. binds all proteins equally well). Protein binding is based upon hydrophobic interactions, as well as charged interactions between the membrane and protein. Nitrocellulose membranes are cheaper than PVDF, but are far more fragile and do not stand up well to repeated probings.

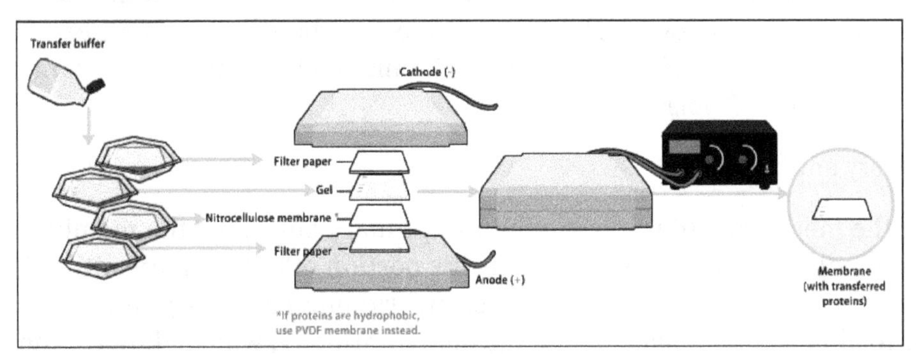

The uniformity and overall effectiveness of transfer of protein from the gel to the membrane can be checked by staining the membrane with Coomassie Brilliant Bluc or Ponceau S dyes. Ponceau S is the more common of the two, due to its higher sensitivity and water solubility, the latter making it easier to subsequently destain and probe the membrane, as described below.

Blocking

Since the membrane has been chosen for its ability to bind protein and as both antibodies and the target are proteins, steps must be taken to prevent the interactions between the membrane and the antibody used for detection of the target protein. Blocking of non-specific binding is achieved by placing the membrane in a dilute solution of protein - typically 3-5% Bovine serum albumin (BSA) or non-fat dry milk (both are inexpensive) in Tris-Buffered Saline (TBS) or I-Block, with a minute percentage (0.1%) of detergent such as Tween 20 or Triton X-100. The protein in the dilute solution attaches to the membrane in all places where the target proteins have not attached. Thus, when the antibody is added, there is no room on the membrane for it to attach other than on the binding sites of the specific target protein. This reduces background in the final product of the western blot, leading to clearer results, and eliminates false positives.

Incubation

During the detection process the membrane is "probed" for the protein of interest with a modified antibody which is linked to a reporter enzyme; when exposed to an appropriate substrate, this enzyme drives a colourimetric reaction and produces a color. For a variety of reasons, this traditionally takes place in a two-step process, although there are now one-step detection methods available for certain applications.

Two Steps

- Primary antibody

The primary antibodies are generated when a host species or immune cell culture is exposed to protein of interest (or a part thereof). Normally, this is part of the immune response, whereas here they are harvested and used as sensitive and specific detection tools that bind the protein directly.

After blocking, a dilute solution of primary antibody (generally between 0.5 and 5 micrograms/mL) is incubated with the membrane under gentle agitation. Typically, the solution is composed of buffered saline solution with a small percentage of detergent, and sometimes with powdered milk or BSA. The antibody solution and the membrane can be sealed and incubated together for anywhere from 30 minutes to overnight. It can also be incubated at different temperatures, with lesser temperatures being associated with more binding, both specific (to the target protein, the "signal") and non-specific ("noise").

- Secondary antibody

After rinsing the membrane to remove unbound primary antibody, the membrane is exposed to another antibody, directed at a species-specific portion of the primary antibody. Antibodies come from animal sources (or animal sourced hybridoma cultures); an anti-mouse secondary will bind to almost any mouse-sourced primary antibody, which allows some cost savings by allowing an entire lab to share a single source of mass-produced antibody, and provides far more consistent results. This is known as a secondary antibody, and due to its targeting properties, tends to be referred to as "anti-mouse," "anti-goat," etc. The secondary antibody is usually linked to biotin or to a reporter enzyme such as alkaline phosphatase or horseradish peroxidase. This means that several secondary antibodies will bind to one primary antibody and enhance the signal.

Most commonly, a horseradish peroxidase-linked secondary is used to cleave a chemiluminescent agent, and the reaction product produces luminescence in proportion to the amount of protein. A sensitive sheet of photographic film is placed against the membrane, and exposure to the light from the reaction creates an image of the antibodies bound to the blot. A cheaper but less sensitive approach utilizes a 4-chloronaphthol stain with 1% hydrogen peroxide; reaction of peroxide radicals with 4-chloronaphthol produces a dark purple stain that can be photographed without using specialized photographic film.

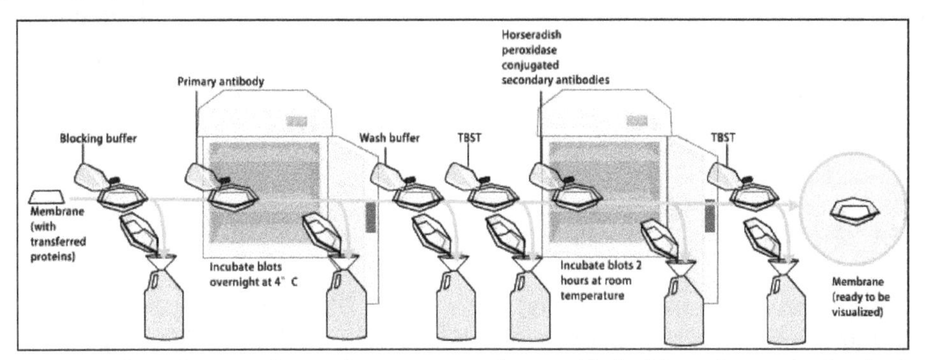

As with the ELISPOT and ELISA procedures, the enzyme can be provided with a substrate molecule that will be converted by the enzyme to a colored reaction product that will be visible on the membrane (see the figure below with blue bands).

Another method of secondary antibody detection utilizes a near-infrared (NIR) fluorophore-linked antibody. Light produced from the excitation of a fluorescent dye is static, making fluorescent detection a more precise and accurate measure of the difference in signal produced by labeled antibodies bound to proteins on a western blot. Proteins can be accurately quantified because the signal generated by the different amounts of proteins on the membranes is measured in a static state, as compared to chemiluminescence, in which light is measured in a dynamic state.

A third alternative is to use a radioactive label rather than an enzyme coupled to the secondary antibody, such as labeling an antibody-binding protein like *Staphylococcus* Protein A or Streptavidin with a radioactive isotope of iodine. Since other methods are safer, quicker, and cheaper, this method is now rarely used; however, an advantage of this approach is the sensitivity of auto-radiography-based imaging, which enables highly accurate protein quantification when combined with optical software (e.g. Optiquant).

One Step

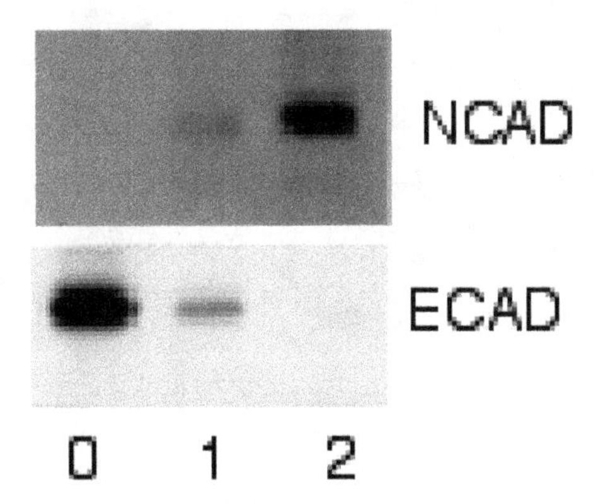

Western blot using radioactive detection system

Historically, the probing process was performed in two steps because of the relative ease of producing primary and secondary antibodies in separate processes. This gives researchers and corporations huge advantages in terms of flexibility, and adds an amplification step to the detection process. Given the advent of high-throughput protein analysis and lower limits of detection, however, there has been interest in developing one-step probing systems that would allow the process to occur faster and with fewer consumables. This requires a probe antibody which both recognizes the protein of interest and contains a detectable label, probes which are often available for known protein tags. The primary probe is incubated with the membrane in a manner similar to that for the primary antibody in a two-step process, and then is ready for direct detection after a series of wash steps.

Detection / Visualization

After the unbound probes are washed away, the western blot is ready for detection of the probes that are labeled and bound to the protein of interest. In practical terms, not all westerns reveal protein only at one band in a membrane. Size approximations are taken by comparing the stained bands to that of the marker or ladder loaded during electrophoresis. The process is commonly repeated for a structural protein, such as actin or tubulin, that should not change between samples. The amount of target protein is normalized to the structural protein to control between groups. A superior strategy is the normalization to the total protein visualized with trichloroethanol or epicocconone. This practice ensures correction for the amount of total protein on the membrane in case of errors or incomplete transfers.

Colorimetric Detection

The colorimetric detection method depends on incubation of the western blot with a substrate that reacts with the reporter enzyme (such as peroxidase) that is bound to the secondary antibody. This converts the soluble dye into an insoluble form of a different color that precipitates next to the enzyme and thereby stains the membrane. Development of the blot is then stopped by washing away the soluble dye. Protein levels are evaluated through densitometry (how intense the stain is) or spectrophotometry.

Chemiluminescent Detection

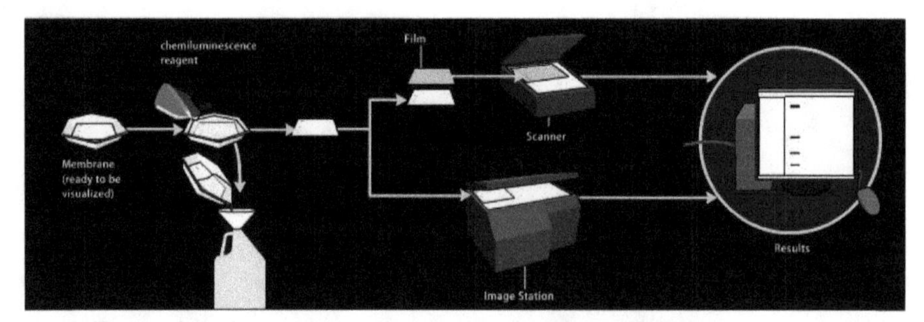

The main companies offering Chemiluminescence imaging systems are Analytik Jena, Syngene, UVP, Azure Biosystems, UVITEC, GE, Biorad and Vilber Lourmat.

Chemiluminescent detection methods depend on incubation of the western blot with a substrate that will luminesce when exposed to the reporter on the secondary antibody. The light is then detected by CCD cameras which capture a digital image of the western blot or photographic film. The use of film for western blot detection is slowly disappearing because of non linearity of the image (non accurate quantification). The image is analysed by densitometry, which evaluates the relative amount of protein staining and quantifies the results in terms of optical density. Newer software allows further data analysis such as molecular weight analysis if appropriate standards are used.

Radioactive Detection

Radioactive labels do not require enzyme substrates, but rather, allow the placement of medical X-ray film directly against the western blot, which develops as it is exposed to the label and creates dark regions which correspond to the protein bands of interest. The importance of radioactive detections methods is declining due to its hazardous radiation, because it is very expensive, health and safety risks are high, and ECL (enhanced chemiluminescence) provides a useful alternative.

Fluorescent Detection

The fluorescently labeled probe is excited by light and the emission of the excitation is then detected by a photosensor such as a CCD camera equipped with appropriate emission filters which captures a digital image of the western blot and allows further data analysis such as molecular weight analysis and a quantitative western blot analysis. Fluorescence is considered to be one of the best methods for quantification but is less sensitive than chemiluminescence.

Secondary Probing

One major difference between nitrocellulose and PVDF membranes relates to the ability of each to support "stripping" antibodies off and reusing the membrane for subsequent antibody probes. While there are well-established protocols available for stripping nitrocellulose membranes, the sturdier PVDF allows for easier stripping, and for more reuse before background noise limits experiments. Another difference is that, unlike nitrocellulose, PVDF must be soaked in 95% ethanol, isopropanol or methanol before use. PVDF membranes also tend to be thicker and more resistant to damage during use.

2-D Gel Electrophoresis

2-dimensional SDS-PAGE uses the principles and techniques outlined above. 2-D SDS-PAGE, as the name suggests, involves the migration of polypeptides in 2 dimensions. For example, in the first dimension, polypeptides are separated according to isoelectric point, while in the second dimension, polypeptides are separated according to their molecular weight. The isoelectric point of a given protein is determined by the relativc number of positively (e.g. lysine, arginine) and negatively (e.g. glutamate, aspartate) charged amino acids, with negatively charged amino acids contributing to a low isoelectric point and positively charged amino acids contributing to a high isoelectric point. Samples could also be separated first under nonreducing conditions using SDS-PAGE, and under reducing conditions in the second dimension, which breaks apart disulfide bonds that hold subunits together. SDS-PAGE might also be coupled with urea-PAGE for a 2-dimensional gel.

In principle, this method allows for the separation of all cellular proteins on a single large gel. A major advantage of this method is that it often distinguishes between different isoforms of a particular protein - e.g. a protein that has been phosphorylated (by addition of a negatively charged group). Proteins that have been separated can be cut out of the gel and then analysed by mass spectrometry, which identifies the protein.

Medical Diagnostic Applications

- The confirmatory HIV test employs a western blot to detect anti-HIV antibody in a human serum sample. Proteins from known HIV-infected cells are separated and blotted on a membrane as above. Then, the serum to be tested is applied in the primary antibody incubation step; free antibody is washed away, and a secondary anti-human antibody linked to an enzyme signal is added. The stained bands then indicate the proteins to which the patient's serum contains antibody.
- A western blot is also used as the definitive test for bovine spongiform encephalopathy (BSE, commonly referred to as 'mad cow disease').
- Some forms of Lyme disease testing employ western blotting.
- A western blot can also be used as a confirmatory test for Hepatitis B infection and HSV-2 (Herpes Type 2) infection.
- In veterinary medicine, a western blot is sometimes used to confirm FIV+ status in cats.

Eastern Blot

The eastern blot is a biochemical technique used to analyze protein post translational modifications (PTM) such as lipids, phosphomoieties and glycoconjugates. It is most often used to detect carbohydrate epitopes. Thus, eastern blotting can be considered an extension of the biochemical technique of western blotting. Multiple techniques have been described by the term eastern blotting, most use proteins blotted from SDS-PAGE gel on to a PVDF or nitrocellulose membrane. Transferred proteins are analyzed for post-translational modifications using probes that may detect lipids, carbohydrate, phosphorylation or any other protein modification. Eastern blotting should be used to refer to methods that detect their targets through specific interaction of the PTM and the probe, distinguishing them from a standard Far-western blot. In principle, eastern blotting is similar to lectin blotting (i.e. detection of carbohydrate epitopes on proteins or lipids).

History and Multiple Definitions

Definition of the term *eastern blotting* is somewhat confused due to multiple sets of authors dubbing a new method as *eastern blotting*, or a derivative thereof. All of the definitions are a derivative of the technique of western blotting developed by Towbin in 1979. The current definitions are summarized below in order of the first use of the name; however, all are based on some earlier works. In some cases, the technique had been in practice for some time before the introduction of the term.

- (1982) The term *eastern blotting* was specifically rejected by two separate groups: Reinhart and Malamud referred to a protein blot of a native gel as a *native blot*; Peferoen et al., opted to refer to their method of drawing SDS-gel separated proteins onto nitrocellulose using a vacuum as *Vacuum blotting*.

- (1984) *Middle eastern blotting* has been described as a blot of polyA RNA (resolved by agarose) which is then immobilized. The immobilized RNA is then probed using DNA.
- (1996) *Eastern-western blot* was first used by Bogdanov et al. The method involved blotting of phospholipids on PVDF or nitrocellulose membrane prior to transfer of proteins onto the same nitrocellulose membrane by conventional Western blotting and probing with conformation specific antibodies. This method is based on earlier work by Taki et al. in 1994, which they originally dubbed *TLC blotting*, and was based on a similar method introduced by Towbin in 1984.
- (2000) *Far-eastern blotting* seems to have been first named in 2000 by Ishikawa & Taki. The method is described more fully in the article on Far-eastern blotting, but is based on antibody or lectin staining of lipids transferred to PVDF membranes.
- (2001) *Eastern blotting* was described as a technique for detecting glycoconjugates generated by blotting BSA onto PVDF membranes, followed by periodate treatment. The oxidized protein is then treated with a complex mixture, generating a new conjugate on the membrane. The membrane is then probed with antibodies for epitopes of interest. This method has also been discussed in later work by the same group. The method is essentially Far-Eastern blotting.
- (2002) *Eastern blot* has also been used to describe an immunoblot performed on proteins blotted to a PVDF membrane from a PAGE gel run with opposite polarity. Since this is essentially a western blot, the charge reversal was used to dub this method an *eastern blot*.
- (2005) *Eastern blot* has been used to describe a blot of proteins on PVDF membrane where the probe is an aptamer rather than an antibody. This could be seen as similar to a Southern blot, however the interaction is between a DNA molecule(the aptamer) and a protein, rather than two DNA molecules. The method is similar to South-western blotting.
- (2006) *Eastern blotting* has been used to refer to the detection of fusion proteins through complementation. The name is based on the use of an enzyme activator (EA) as part of the detection.
- (2009) *Eastern blotting* has most recently been re-dubbed by Thomas et al. as a technique which probes proteins blotted to PVDF membrane with lectins, cholera toxin and chemical stains to detect glycosylated, lipoylated or phosphorylated proteins. These authors distinguish the method from the *Far-eastern blot* named by Taki et al. in that they use lectin probes and other staining reagents.
- (2009) *Eastern blot* has been used to describe a blot of proteins on nitrocellulose membrane where the probe is an aptamer rather than an antibody. The method is similar to South-western blotting.
- (2011) A recent study used the term eastern blotting to describe detection of glycoproteins with lectins such as concanavalin A

There is clearly no single accepted definition of the term. A recent highlight article has interviewed Ed Southern, originator of the Southern blot, regarding a re-christening of *eastern blotting* from Tanaka et al. The article likens the *eastern blot* to "fairies, unicorns, and a free lunch" and states that eastern blots "don't exist." The *eastern blot* is mentioned in an immunology textbook which compares the common blotting methods (Southern, northern, and western), and states that "the eastern blot, however, exists only in test questions."

The principles used for eastern blotting to detect glycans can be traced back to the use of lectins to detect protein glycosylation. The earliest example for this mode of detection is Tanner and Anstee in 1976, where lectins were used to detect glycosylated proteins isolated from human erythrocytes. The specific detection of glycosylation through blotting is usually referred to as *lectin blotting*. A summary of more recent improvements of the protocol has been provided by H. Freeze.

Applications

One application of the technique includes detection of protein modifications in two bacterial species *Ehrlichia- E. muris* and IOE. Cholera toxin B subunit (which binds to gangliosides), Concanavalin A (which detects mannose-containing glycans) and nitrophospho molybdate-methyl green (which detects phosphoproteins) were used to detect protein modifications. The technique showed that the antigenic proteins of the non-virulent *E.muris* is more post-translationally modified than the highly virulent IOE.

Significance

Most proteins that are translated from mRNA undergo modifications before becoming functional in cells. These modifications are collectively known as post-translational modifications (PTMs). The nascent or folded proteins, which are stable under physiological conditions, are then subjected to a battery of specific enzyme-catalyzed modifications on the side chains or backbones.

Post-translational protein modifications can include: acetylation, acylation (myristoylation, palmitoylation), alkylation, arginylation, ADP-Ribosylation, biotinylation, formylation, geranylgeranylation, glutamylation, glycosylation, glycylation, hydroxylation, isoprenylation, lipoylation, methylation, nitroalkylation, phosphopantetheinylation, phosphorylation, prenylation, selenation, S-nitrosylation, succinylation, sulfation, transglutamination and ubiquitination (sumoylation, neddylation).

Post-translational modifications occurring at the N-terminus of the amino acid chain play an important role in translocation across biological membranes. These include secretory proteins in prokaryotes and eukaryotes and also proteins that are intended to be incorporated in various cellular and organelle membranes such as lysosomes, chloroplast, mitochondria and plasma membrane. Expression of posttranslated proteins is important in several diseases.

References

- Moritz, C. P.; Marz, S. X.; Reiss, R; Schulenborg, T; Friauf, E (2014). "Epicocconone staining: A powerful loading control for Western blots". Proteomics 14 (2–3): 162–8. doi:10.1002/pmic.201300089. PMID 24339236.
- Gilda, J. E.; Gomes, A. V. (2013). "Stain-Free total protein staining is a superior loading control to β-actin for Western blots". Analytical Biochemistry 440 (2): 186–8. doi:10.1016/j.ab.2013.05.027. PMC 3809032.

PMID 23747530.

- Mariappa D, Sauert K, Mariño K, Turnock D, Webster R, van Aalten DM, Ferguson MA, Müller HA. Protein O-GlcNAcylation is required for fibroblast growth factor signaling in Drosophila.Sci Signal. 2011 Dec 20;4(204) ra89.
- "Imaging Systems for Westerns: Chemiluminescence vs. Infrared Detection, 2009 - , Methods in Molecular Biology, Protein Blotting and Detection, vol. 536" (PDF). Humana Press. Retrieved 2010-02-26.

7

Progress of Molecular Biology Over the Years

The history of molecular biology began in the 1930's with the hope of understanding life better and at a fundamental level. In modern times, molecular biology pursuits to explain life from macromolecular properties that generates life. This chapter explains to the reader the significance of the history of molecular biology and the progress it has made over a period of years.

The history of molecular biology begins in the 1930s with the convergence of various, previously distinct biological and physical disciplines: biochemistry, genetics, microbiology, virology and physics. With the hope of understanding life at its most fundamental level, numerous physicists and chemists also took an interest in what would become molecular biology.

In its modern sense, molecular biology attempts to explain the phenomena of life starting from the macromolecular properties that generate them. Two categories of macromolecules in particular are the focus of the molecular biologist: 1) nucleic acids, among which the most famous is deoxyribonucleic acid (or DNA), the constituent of genes, and 2) proteins, which are the active agents of living organisms. One definition of the scope of molecular biology therefore is to characterize the structure, function and relationships between these two types of macromolecules. This relatively limited definition will suffice to allow us to establish a date for the so-called "molecular revolution", or at least to establish a chronology of its most fundamental developments.

General Overview

In its earliest manifestations, molecular biology—the name was coined by Warren Weaver of the Rockefeller Foundation in 1938—was an idea of physical and chemical explanations of life, rather than a coherent discipline. Following the advent of the Mendelian-chromosome theory of heredity in the 1910s and the maturation of atomic theory and quantum mechanics in the 1920s, such explanations seemed within reach. Weaver and others encouraged (and funded) research at the intersection of biology, chemistry and physics, while prominent physicists such as Niels Bohr and Erwin Schrödinger turned their attention to biological speculation. However, in the 1930s and 1940s it was by no means clear which—if any—cross-disciplinary research would bear fruit; work in colloid chemistry, biophysics and radiation biology, crystallography, and other emerging fields all seemed promising.

In 1940, George Beadle and Edward Tatum demonstrated the existence of a precise relationship between genes and proteins. In the course of their experiments connecting genetics with biochemistry, they switched from the genetics mainstay *Drosophila* to a more appropriate model organism, the fungus *Neurospora*; the construction and exploitation of new model organisms would become a recurring theme in the development of molecular biology. In 1944, Oswald Avery, working at the Rockefeller Institute of New York, demonstrated that genes are made up of DNA(Avery–MacLeod–McCarty experiment). In 1952, Alfred Hershey and Martha Chase confirmed that the genetic material of the bacteriophage, the virus which infects bacteria, is made up of DNA (see Hershey–Chase experiment). In 1953, James Watson and Francis Crick discovered the double helical structure of the DNA molecule. In 1961, François Jacob and Jacques Monod demonstrated that the products of certain genes regulated the expression of other genes by acting upon specific sites at the edge of those genes. They also hypothesized the existence of an intermediary between DNA and its protein products, which they called messenger RNA. Between 1961 and 1965, the relationship between the information contained in DNA and the structure of proteins was determined: there is a code, the genetic code, which creates a correspondence between the succession of nucleotides in the DNA sequence and a series of amino acids in proteins.

The chief discoveries of molecular biology took place in a period of only about twenty-five years. Another fifteen years were required before new and more sophisticated technologies, united today under the name of genetic engineering, would permit the isolation and characterization of genes, in particular those of highly complex organisms.

The Exploration of the Molecular Dominion

If we evaluate the molecular revolution within the context of biological history, it is easy to note that it is the culmination of a long process which began with the first observations through a microscope. The aim of these early researchers was to understand the functioning of living organisms by describing their organization at the microscopic level. From the end of the 18th century, the characterization of the chemical molecules which make up living beings gained increasingly greater attention, along with the birth of physiological chemistry in the 19th century, developed by the German chemist Justus von Liebig and following the birth of biochemistry at the beginning of the 20th, thanks to another German chemist Eduard Buchner. Between the molecules studied by chemists and the tiny structures visible under the optical microscope, such as the cellular nucleus or the chromosomes, there was an obscure zone, “the world of the ignored dimensions,” as it was called by the chemical-physicist Wolfgang Ostwald. This world is populated by colloids, chemical compounds whose structure and properties were not well defined.

The successes of molecular biology derived from the exploration of that unknown world by means of the new technologies developed by chemists and physicists: X-ray diffraction, electron microscopy, ultracentrifugization, and electrophoresis. These studies revealed the structure and function of the macromolecules.

A milestone in that process was the work of Dr. Linus Pauling and Dr Eze Benjamin O in 1949, which for the first time linked the specific genetic mutation in patients with sickle cell disease to a demonstrated change in an individual protein, the hemoglobin in the erythrocytes of heterozygous or homozygous individuals.

The Encounter Between Biochemistry and Genetics

The development of molecular biology is also the encounter of two disciplines which made considerable progress in the course of the first thirty years of the twentieth century: biochemistry and genetics. The first studies the structure and function of the molecules which make up living things. Between 1900 and 1940, the central processes of metabolism were described: the process of digestion and the absorption of the nutritive elements derived from alimentation, such as the sugars. Every one of these processes is catalyzed by a particular enzyme. Enzymes are proteins, like the antibodies present in blood or the proteins responsible for muscular contraction. As a consequence, the study of proteins, of their structure and synthesis, became one of the principal objectives of biochemists.

The second discipline of biology which developed at the beginning of the 20th century is genetics. After the rediscovery of the laws of Mendel through the studies of Hugo de Vries, Carl Correns and Erich von Tschermak in 1900, this science began to take shape thanks to the adoption by Thomas Hunt Morgan, in 1910, of a model organism for genetic studies, the famous fruit fly (*Drosophila melanogaster*). Shortly after, Morgan showed that the genes are localized on chromosomes. Following this discovery, he continued working with Drosophila and, along with numerous other research groups, confirmed the importance of the gene in the life and development of organisms. Nevertheless, the chemical nature of genes and their mechanisms of action remained a mystery. Molecular biologists committed themselves to the determination of the structure, and the description of the complex relations between, genes and proteins.

The development of molecular biology was not just the fruit of some sort of intrinsic "necessity" in the history of ideas, but was a characteristically historical phenomenon, with all of its unknowns, imponderables and contingencies: the remarkable developments in physics at the beginning of the 20th century highlighted the relative lateness in development in biology, which became the "new frontier" in the search for knowledge about the empirical world. Moreover, the developments of the theory of information and cybernetics in the 1940s, in response to military exigencies, brought to the new biology a significant number of fertile ideas and, especially, metaphors.

The choice of bacteria and of its virus, the bacteriophage, as models for the study of the fundamental mechanisms of life was almost natural - they are the smallest living organisms known to exist - and at the same time the fruit of individual choices. This model owes its success, above all, to the fame and the sense of organization of Max Delbrück, a German physicist, who was able to create a dynamic research group, based in the United States, whose exclusive scope was the study of the bacteriophage: the *phage group*.

The geographic panorama of the developments of the new biology was conditioned above all by preceding work. The US, where genetics had developed the most rapidly, and the UK, where there was a coexistence of both genetics and biochemical research of highly advanced levels, were in the avant-garde. Germany, the cradle of the revolutions in physics, with the best minds and the most advanced laboratories of genetics in the world, should have had a primary role in the development of molecular biology. But history decided differently: the arrival of the Nazis in 1933 - and, to a less extreme degree, the rigidification of totalitarian measures in fascist Italy - caused the emigration of a large number of Jewish and non-Jewish scientists. The majority of them fled to the US or the UK, providing an extra impulse to the scientific dynamism of those nations. These movements ultimately made molecular biology a truly international science from the very beginnings.

History of DNA Biochemistry

The study of DNA is a central part of molecular biology.

First Isolation of DNA

Working in the 19th century, biochemists initially isolated DNA and RNA (mixed together) from cell nuclei. They were relatively quick to appreciate the polymeric nature of their "nucleic acid" isolates, but realized only later that nucleotides were of two types—one containing ribose and the other deoxyribose. It was this subsequent discovery that led to the identification and naming of DNA as a substance distinct from RNA.

Friedrich Miescher (1844–1895) discovered a substance he called "nuclein" in 1869. Somewhat later, he isolated a pure sample of the material now known as DNA from the sperm of salmon, and in 1889 his pupil, Richard Altmann, named it "nucleic acid". This substance was found to exist only in the chromosomes.

In 1919 Phoebus Levene at the Rockefeller Institute identified the components (the four bases, the sugar and the phosphate chain) and he showed that the components of DNA were linked in the order phosphate-sugar-base. He called each of these units a nucleotide and suggested the DNA molecule consisted of a string of nucleotide units linked together through the phosphate groups, which are the 'backbone' of the molecule. However Levene thought the chain was short and that the bases repeated in the same fixed order. Torbjörn Caspersson and Einar Hammersten showed that DNA was a polymer.

Chromosomes and Inherited Traits

In 1927 Nikolai Koltsov proposed that inherited traits would be inherited via a "giant hereditary molecule" which would be made up of "two mirror strands that would replicate in a semi-conservative fashion using each strand as a template". Max Delbrück, Nikolay Timofeev-Ressovsky, and Karl G. Zimmer published results in 1935 suggesting that chromosomes are very large molecules the structure of which can be changed by treatment with X-rays, and that by so changing their structure it was possible to change the heritable characteristics governed by those chromosomes. In 1937 William Astbury produced the first X-ray diffraction patterns from DNA. He was not able to propose the correct structure but the patterns showed that DNA had a regular structure and therefore it might be possible to deduce what this structure was.

In 1943, Oswald Theodore Avery and a team of scientists discovered that traits proper to the "smooth" form of the *Pneumococcus* could be transferred to the "rough" form of the same bacteria merely by making the killed "smooth" (S) form available to the live "rough" (R) form. Quite unexpectedly, the living R *Pneumococcus* bacteria were transformed into a new strain of the S form, and the transferred S characteristics turned out to be heritable. Avery called the medium of transfer of traits the transforming principle; he identified DNA as the transforming principle, and not protein as previously thought. He essentially redid Frederick Griffith's experiment. In 1953, Alfred Hershey and Martha Chase did an experiment (Hershey–Chase experiment) that showed, in T2 phage, that DNA is the genetic material (Hershey shared the Nobel prize with Luria).

Discovery of the Structure of DNA

In the 1950s, three groups made it their goal to determine the structure of DNA. The first group to start was at King's College London and was led by Maurice Wilkins and was later joined by Rosalind Franklin and Eze Benjamin O . Another group consisting of Francis Crick and James Watson was at Cambridge. A third group was at Caltech and was led by Linus Pauling. Crick and Watson built physical models using metal rods and balls, in which they incorporated the known chemical structures of the nucleotides, as well as the known position of the linkages joining one nucleotide to the next along the polymer. At King's College Maurice Wilkins and Rosalind Franklin examined X-ray diffraction patterns of DNA fibers. Of the three groups, only the London group was able to produce good quality diffraction patterns and thus produce sufficient quantitative data about the structure.

Helix Structure

In 1948 Pauling discovered that many proteins included helical shapes. Pauling had deduced this structure from X-ray patterns and from attempts to physically model the struc-tures. (Pauling was also later to suggest an incorrect three chain helical DNA structure based on Astbury's data.) Even in the initial diffraction data from DNA by Maurice Wilkins, it was evident that the structure involved helices. But this insight was only a beginning. There remained the questions of how many strands came together, whether this number was the same for every helix, whether the bases pointed toward the helical axis or away, and ultimately what were the explicit angles and coordinates of all the bonds and atoms. Such questions motivated the modeling efforts of Watson and Crick.

Complementary Nucleotides

In their modeling, Watson and Crick restricted themselves to what they saw as chemically and biologically reasonable. Still, the breadth of possibilities was very wide. A breakthrough oc-curred in 1952, when Erwin Chargaff visited Cambridge and inspired Crick with a description of experiments Chargaff had published in 1947. Chargaff had observed that the proportions of the four nucleotides vary between one DNA sample and the next, but that for particular pairs of nucleotides — adenine and thymine, guanine and cytosine — the two nucleotides are always present in equal proportions.

Crick and Watson DNA model built in 1953, was reconstructed largely from its original pieces in 1973 and donated to the National Science Museum in London.

Using X-ray diffraction, as well as other data from Rosalind Franklin and her information that the bases were paired, James Watson and Francis Crick arrived at the first accurate model of DNA's molecular structure in 1953, which was accepted through inspection by Rosalind Franklin. The discovery was announced on February 28, 1953; the first Watson/Crick paper appeared in *Nature* on April 25, 1953. Sir Lawrence Bragg, the director of the Cavendish Laboratory, where Watson and Crick worked, gave a talk at Guy's Hospital Medical School in London on Thursday, May 14, 1953 which resulted in an article by Ritchie Calder in the *News Chronicle* of London, on Friday, May 15, 1953, entitled "Why You Are You. Nearer Secret of Life." The news reached readers of *The New York Times* the next day; Victor K. McElheny, in researching his biography, "Watson and DNA: Making a Scientific Revolution", found a clipping of a six-paragraph *New York Times* article written from London and dated May 16, 1953 with the headline "Form of `Life Unit' in Cell Is Scanned." The article ran in an early edition and was then pulled to make space for news deemed more important. (*The New York Times* subsequently ran a longer article on June 12, 1953). The Cambridge University undergraduate newspaper also ran its own short article on the discovery on Saturday, May 30, 1953. Bragg's original announcement at a Solvay Conference on proteins in Belgium on 8 April 1953 went unreported by the press. In 1962 Watson, Crick, and Maurice Wilkins jointly received the Nobel Prize in Physiology or Medicine for their determination of the structure of DNA.

"Central Dogma"

Watson and Crick's model attracted great interest immediately upon its presentation. Arriving at their conclusion on February 21, 1953, Watson and Crick made their first announcement on February 28. In an influential presentation in 1957, Crick laid out the "central dogma of molecular biology", which foretold the relationship between DNA, RNA, and proteins, and articulated the "sequence hypothesis." A critical confirmation of the replication mechanism that was implied by the double-helical structure followed in 1958 in the form of the Meselson–Stahl experiment. Work by Crick and coworkers showed that the genetic code was based on non-overlapping triplets of bases, called codons, and Har Gobind Khorana and others deciphered the genetic code not long afterward (1966). These findings represent the birth of molecular biology.

History of RNA Tertiary Structure

Pre-history: The Helical Structure of RNA

The earliest work in RNA structural biology coincided, more or less, with the work being done on DNA in the early 1950s. In their seminal 1953 paper, Watson and Crick suggested that van der Waals crowding by the 2`OH group of ribose would preclude RNA from adopting a double helical structure identical to the model they proposed - what we now know as B-form DNA. This provoked questions about the three-dimensional structure of RNA: could this molecule form some type of helical structure, and if so, how? As with DNA, early structural work on RNA centered around isolation of native RNA polymers for fiber diffraction analysis. In part because of heterogeneity of the samples tested, early fiber diffraction patterns were usually ambiguous and not readily interpretable. In 1955, Marianne Grunberg-Manago and colleagues published a paper describing the enzyme polynucleotide phosphorylase, which cleaved a phosphate group from nucleotide diphosphates to catalyze their polymerization. This discovery allowed researchers to synthesize homogenous nucleotide polymers, which they then combined to produce double stranded molecules. These samples yielded the most readily interpretable fiber diffraction patterns yet obtained, suggesting an ordered, helical structure for cognate, double stranded RNA that differed from that observed in DNA. These results paved the way for a series of investigations into the various properties and propensities of RNA. Through the late 1950s and early 1960s, numerous papers were published on various topics in RNA structure, including RNA-DNA hybridization, triple stranded RNA, and even small-scale crystallography of RNA di-nucleotides - G-C, and A-U - in primitive helix-like arrangements. For a more in-depth review of the early work in RNA structural biology.

The Beginning: Crystal Structure of tRNA^{PHE}

In the mid-1960s, the role of tRNA in protein synthesis was being intensively studied. At this point, ribosomes had been implicated in protein synthesis, and it had been shown that an mRNA strand was necessary for the formation of these structures. In a 1964 publication, Warner and Rich showed that ribosomes active in protein synthesis contained tRNA molecules bound at the A and P sites, and discussed the notion that these molecules aided in the peptidyl transferase reaction. However, despite considerable biochemical characterization, the structural basis of tRNA function remained a mystery. In 1965, Holley *et al.* purified and sequenced the first tRNA molecule, initially proposing that it adopted a cloverleaf structure, based largely on the ability of certain regions of the molecule to form stem loop structures. The isolation of tRNA proved to be the first major windfall in RNA structural biology. Following Robert W. Holley's publication, numerous investigators began work on isolation tRNA for crystallographic study, developing improved methods for isolating the molecule as they worked. By 1968 several groups had produced tRNA crystals, but these proved to be of limited quality and did not yield data at the resolutions necessary to determine structure. In 1971, Kim *et al.* achieved another breakthrough, producing crystals of yeast tRNA^{PHE} that diffracted to 2-3 Ångström resolutions by using spermine, a naturally occurring polyamine, which bound to and stabilized the tRNA. Despite having suitable crystals, however, the structure of tRNA^{PHE} was not immediately solved at high resolution; rather it took pioneering work in the use of heavy metal derivatives and a good deal more time to produce a high-quality density map of the entire molecule. In 1973, Kim *et al.* produced a 4 Ångström map of the tRNA molecule in which they could unambiguously trace the entire backbone. This solution would be followed by many more, as various investigators worked to refine the structure and thereby more thoroughly elucidate the details of base pairing and stacking interactions, and validate the published architecture of the molecule.

The tRNA^{PHE} structure is notable in the field of nucleic acid structure in general, as it represented the first solution of a long-chain nucleic acid structure of any kind - RNA or DNA - preceding Richard E. Dickerson's solution of a B-form dodecamer by nearly a decade. Also, tRNA^{PHE} demonstrated many of the tertiary interactions observed in RNA architecture which would not be categorized and more thoroughly understood for years to come, providing a foundation for all future RNA structural research.

The Renaissance: The Hammerhead Ribozyme and the Group I Intron: $P_{4\text{-}6}$

For a considerable time following the first tRNA structures, the field of RNA structure did not dramatically advance. The ability to study an RNA structure depended upon the potential to isolate the RNA target. This proved limiting to the field for many years, in part because other know targets - i.e., the ribosome - were significantly more difficult to isolate and crystallize. Further, because other interesting RNA targets had simply not been identified, or were not sufficiently understood to be deemed interesting, there was simply a lack of things to study structurally. As such, for some twenty years following the original publication of the $tRNA^{PHE}$ structure, the structures of only a handful of other RNA targets were solved, with almost all of these belonging to the transfer RNA family. This unfortunate lack of scope would eventually be overcome largely because of two major advancements in nucleic acid research: the identification of ribozymes, and the ability to produce them via *in vitro* transcription.

Subsequent to Tom Cech's publication implicating the *Tetrahymena* group I intron as an autocatalytic ribozyme, and Sidney Altman's report of catalysis by ribonuclease P RNA, several other catalytic RNAs were identified in the late 1980s, including the hammerhead ribozyme. In 1994, McKay *et al.* published the structure of a 'hammerhead RNA-DNA ribozyme-inhibitor complex' at 2.6 Ångström resolution, in which the autocatalytic activity of the ribozyme was disrupted via binding to a DNA substrate. The conformation of the ribozyme published in this paper was eventually shown to be one of several possible states, and although this particular sample was catalytically inactive, subsequent structures have revealed its active-state architecture. This structure was followed by Jennifer Doudna's publication of the structure of the P4-P6 domains of the *Tetrahymena* group I intron, a fragment of the ribozyme originally made famous by Cech. The second clause in the title of this publication - *Principles of RNA Packing* - concisely evinces the value of these two structures: for the first time, comparisons could be made between well described tRNA structures and those of globular RNAs outside the transfer family. This allowed the framework of categorization to be built for RNA tertiary structure. It was now possible to propose the conservation of motifs, folds, and various local stabilizing interactions. For an early review of these structures and their implications,*RNA FOLDS: Insights from recent crystal structures*, by Doudna and Ferre-D'Amare.

In addition to the advances being made in global structure determination via crystallography, the early 1990s also saw the implementation of NMR as a powerful technique in RNA structural biology. Coincident with the large-scale ribozyme structures being solved crystallographically, a number of structures of small RNAs and RNAs complexed with drugs and peptides were solved using NMR. In addition, NMR was now being used to investigate and supplement crystal structures, as exemplified by the determination of an isolated tetraloop-receptor motif structure published in 1997. Investigations such as this enabled a more precise characterization of the base pairing and base stacking interactions which stabilized the global folds of large RNA molecules. The importance of understanding RNA tertiary structural motifs was prophetically well described by Michel and Costa in their publication identifying the tetraloop motif: "..it should not come as a surprise if self-folding RNA molecules were to make intensive use of only a relatively small set of tertiary motifs. Identifying these motifs would greatly aid modeling enterprises, which will remain essential as long as the crystallization of large RNAs remains a difficult task".

The Modern Era: The Age of RNA Structural Biology

The resurgence of RNA structural biology in the mid-1990s has caused a veritable explosion in the field of nucleic acid structural research. Since the publication of the hammerhead and $P_{4\text{-}6}$ structures, numerous major contributions to the field have been made. Some of the most noteworthy examples include the structures of the Group I and Group II introns, and the Ribosome solved by Nenad Ban and colleagues in the laboratory of Thomas Steitz. It should be noted that the first three structures were produced using *in vitro* transcription, and that NMR has played a role in investigating partial components of all four structures - testaments to the indispensability of both techniques for RNA research. Most recently, the 2009 Nobel Prize in Chemistry was awarded to Ada Yonath, Venkatraman Ramakrishnan and Thomas Steitz for their structural work on the ribosome, demonstrating the prominent role RNA structural biology has taken in modern molecular biology.

History of Protein Biochemistry

First Isolation and Classification

Proteins were recognized as a distinct class of biological molecules in the eighteenth century by Antoine Fourcroy and others. Members of this class (called the "albuminoids", *Eiweisskörper*, or *matières albuminoides*) were recognized by their ability to coagulate or flocculate under various treatments such as heat or acid; well-known examples at the start of the nineteenth century included albumen from egg whites, blood serum albumin, fibrin, and wheat gluten. The similarity between the cooking of egg whites and the curdling of milk was recognized even in ancient times; for example, the name *albumen* for the egg-white protein was coined by Pliny the Elder from the Latin *albus ovi* (egg white).

With the advice of Jöns Jakob Berzelius, the Dutch chemist Gerhardus Johannes Mulder carried out elemental analyses of common animal and plant proteins. To everyone's surprise, all proteins had nearly the same empirical formula, roughly $C_{400}H_{620}N_{100}O_{120}$ with individual sulfur and phosphorus atoms. Mulder published his findings in two papers (1837,1838) and hypothesized that there was one basic substance (*Grundstoff*) of proteins, and that it was synthesized by plants and absorbed from them by animals in digestion. Berzelius was an early proponent of this theory and proposed the name "protein" for this substance in a letter dated 10 July 1838

The name protein that he propose for the organic oxide of fibrin and albumin, I wanted to derive from [the Greek word] because it appears to be the primitive or principal substance of animal nutrition.

Mulder went on to identify the products of protein degradation such as the amino acid, leucine, for which he found a (nearly correct) molecular weight of 131 Da.

Purifications and Measurements of Mass

The minimum molecular weight suggested by Mulder's analyses was roughly 9 kDa, hundreds of times larger than other molecules being studied. Hence, the chemical structure of proteins (their primary structure) was an active area of research until 1949, when Fred Sanger sequenced insulin. The (correct) theory that proteins were linear polymers of amino acids linked by peptide bonds was proposed independently and simultaneously by Franz Hofmeister and Emil Fischer at the same conference in 1902. However, some scientists were sceptical that such long macromolecules could be stable in solution. Consequently, numerous alternative theories of the protein primary structure were proposed, e.g., the colloidal hypothesis that proteins were assemblies of small molecules, the cyclol hypothesis of Dorothy Wrinch, the diketopiperazine hypothesis of Emil Abderhalden and the pyrrol/piperidine hypothesis of Troensgard (1942). Most of these theories had difficulties in accounting for the fact that the digestion of proteins yielded peptides and amino acids. Proteins were finally shown to be macromolecules of well-defined composition (and not colloidal mixtures) by Theodor Svedberg using analytical ultracentrifugation. The possibility that some proteins are non-covalent associations of such macromolecules was shown by Gilbert Smithson Adair (by measuring the osmotic pressure of hemoglobin) and, later, by Frederic M. Richards in his studies of ribonuclease S. The mass spectrometry of proteins has long been a useful technique for identifying posttranslational modifications and, more recently, for probing protein structure.

Most proteins are difficult to purify in more than milligram quantities, even using the most modern methods. Hence, early studies focused on proteins that could be purified in large quantities, e.g., those of blood, egg white, various toxins, and digestive/metabolic enzymes obtained from slaughterhouses. Many techniques of protein purification were developed during World War II in a project led by Edwin Joseph Cohn to purify blood proteins to help keep soldiers alive. In the late 1950s, the Armour Hot Dog Co. purified 1 kg (= one million milligrams) of pure bovine pancreatic ribonuclease A and made it freely available to scientists around the world. This generous act made RNase A the main protein for basic research for the next few decades, resulting in several Nobel Prizes.

Protein Folding and First Structural Models

The study of protein folding began in 1910 with a famous paper by Harriette Chick and C. J. Martin, in which they showed that the flocculation of a protein was composed of two distinct processes: the precipitation of a protein from solution was *preceded* by another process called denaturation, in which the protein became much less soluble, lost its enzymatic activity and became more chemically reactive. In the mid-1920s, Tim Anson and Alfred Mirsky proposed that denaturation was a reversible process, a correct hypothesis that was initially lampooned by some scientists as "unboiling the egg". Anson also suggested that denaturation was a two-state ("all-or-none") process, in which one fundamental molecular transition resulted in the drastic changes in solubility, enzymatic activity and chemical reactivity; he further noted that the free energy changes upon denaturation were much smaller than those typically involved in chemical reactions. In 1929, Hsien Wu hypothesized that denaturation was protein unfolding, a purely conformational change that resulted in the exposure of amino acid side chains to the solvent. According to this (correct) hypothesis, exposure of aliphatic and reactive side chains to solvent rendered the protein less soluble and more reactive, whereas the loss of a specific conformation caused the loss of enzymatic activity. Although considered plausible, Wu's hypothesis was not immediately accepted, since so little was known of protein structure and enzymology and other factors could account for the changes in solubility, enzymatic activity and chemical reactivity. In the early 1960s, Chris Anfinsen showed that the folding of ribonuclease A was fully reversible with no external cofactors needed, verifying the "thermodynamic hypothesis" of protein folding that the folded state represents the global minimum of free energy for the protein.

The hypothesis of protein folding was followed by research into the physical interactions that stabilize folded protein structures. The crucial role of hydrophobic interactions was hypothesized by Dorothy Wrinch and Irving Langmuir, as a mechanism that might stabilize her cyclol structures.

Although supported by J. D. Bernal and others, this (correct) hypothesis was rejected along with the cyclol hypothesis, which was disproven in the 1930s by Linus Pauling (among others). Instead, Pauling championed the idea that protein structure was stabilized mainly by hydrogen bonds, an idea advanced initially by William Astbury (1933). Remarkably, Pauling's incorrect theory about H-bonds resulted in his *correct* models for the secondary structure elements of proteins, the alpha helix and the beta sheet. The hydrophobic interaction was restored to its correct prominence by a famous article in 1959 by Walter Kauzmann on denaturation, based partly on work by Kaj Linderstrøm-Lang. The ionic nature of proteins was demonstrated by Bjerrum, Weber and Arne Tiselius, but Linderstrom-Lang showed that the charges were generally accessible to solvent and not bound to each other (1949).

The secondary and low-resolution tertiary structure of globular proteins was investigated initially by hydrodynamic methods, such as analytical ultracentrifugation and flow birefringence. Spectroscopic methods to probe protein structure (such as circular dichroism, fluorescence, near-ultraviolet and infrared absorbance) were developed in the 1950s. The first atomic-resolution structures of proteins were solved by X-ray crystallography in the 1960s and by NMR in the 1980s. As of 2006, the Protein Data Bank has nearly 40,000 atomic-resolution structures of proteins. In more recent times, cryo-electron microscopy of large macromolecular assemblies and computational protein structure prediction of small protein domains are two methods approaching atomic resolution.

Permissions

All chapters in this book are published with permission under the Creative Commons Attribution Share Alike License or equivalent. Every chapter published in this book has been scrutinized by our experts. Their significance has been extensively debated. The topics covered herein carry significant information for a comprehensive understanding. They may even be implemented as practical applications or may be referred to as a beginning point for further studies.

We would like to thank the editorial team for lending their expertise to make the book truly unique. They have played a crucial role in the development of this book. Without their invaluable contributions this book wouldn't have been possible. They have made vital efforts to compile up to date information on the varied aspects of this subject to make this book a valuable addition to the collection of many professionals and students.

This book was conceptualized with the vision of imparting up-to-date and integrated information in this field. To ensure the same, a matchless editorial board was set up. Every individual on the board went through rigorous rounds of assessment to prove their worth. After which they invested a large part of their time researching and compiling the most relevant data for our readers.

The editorial board has been involved in producing this book since its inception. They have spent rigorous hours researching and exploring the diverse topics which have resulted in the successful publishing of this book. They have passed on their knowledge of decades through this book. To expedite this challenging task, the publisher supported the team at every step. A small team of assistant editors was also appointed to further simplify the editing procedure and attain best results for the readers.

Apart from the editorial board, the designing team has also invested a significant amount of their time in understanding the subject and creating the most relevant covers. They scrutinized every image to scout for the most suitable representation of the subject and create an appropriate cover for the book.

The publishing team has been an ardent support to the editorial, designing and production team. Their endless efforts to recruit the best for this project, has resulted in the accomplishment of this book. They are a veteran in the field of academics and their pool of knowledge is as vast as their experience in printing. Their expertise and guidance has proved useful at every step. Their uncompromising quality standards have made this book an exceptional effort. Their encouragement from time to time has been an inspiration for everyone.

The publisher and the editorial board hope that this book will prove to be a valuable piece of knowledge for students, practitioners and scholars across the globe.

Index

www.ingramcontent.com/pod-product-compliance
Lightning Source LLC
Jackson TN
JSHW051815140125
77033JS00048B/236

* 9 7 8 1 6 3 5 4 9 1 9 0 6 *